PENGUIN BOOKS

THE GOLDILOCKS ENIGMA

Paul Davies is an internationally acclaimed physicist, cosmologist and writer. He is the author of over twenty books, including *The Mind of God*, *About Time*, *The Origin of Life* and *How to Build a Time Machine*.

Davies's talents as a communicator of science have been recognized in the UK by the 2001 Kelvin Medal and Prize awarded by the Institute of Physics, and the 2002 Faraday Prize of the Royal Society. In Australia he has received an Advance Australia Award and two Eureka Prizes. For his contributions to the deeper implications of science, Davies received the Templeton Prize in 1995.

Paul Davies obtained a PhD from the University of London and held academic appointments at the universities of Cambridge, Newcastle upon Tyne and Adelaide, before helping to establish the Australian Centre for Astrobiology at Macquarie University in Sydney. He currently lives in the United States and works at Arizona State University.

D0191472

PAUL DAVIES

The Goldilocks Enigma

WHY IS THE UNIVERSE JUST
RIGHT FOR LIFE?

PENGUIN BOOKS

PENGUIN BOOKS

Published by the Penguin Group
Penguin Books Ltd, 80 Strand, London WC2R 0RL, England
Penguin Group (USA) Inc., 375 Hudson Street, New York, New York 10014, USA
Penguin Group (Canada), 90 Eglinton Avenue East, Suite 700, Toronto, Ontario, Canada M4P 2Y3
(a division of Pearson Penguin Canada Inc.)
Penguin Ireland, 25 St Stephen's Green, Dublin 2, Ireland
(a division of Penguin Books Ltd)
Penguin Group (Australia), 250 Camberwell Road, Camberwell, Victoria 3124, Australia
(a division of Pearson Australia Group Pty Ltd)
Penguin Books India Pvt Ltd, 11 Community Centre, Panchsheel Park, New Delhi – 110 017, India
Penguin Group (NZ), 67 Apollo Drive, Rosedale, North Shore 0632, New Zealand
(a division of Pearson New Zealand Ltd)
Penguin Books (South Africa) (Pty) Ltd, 24 Sturdee Avenue, Rosebank, Johannesburg 2196, South Africa

Penguin Books Ltd, Registered Offices: 80 Strand, London WC2R 0RL, England

www.penguin.com

First published by Allen Lane 2006
Published in Penguin Books 2007
5

Copyright © Paul Davies, 2006
All rights reserved

The moral right of the author has been asserted

Typeset by Rowland Phototypesetting Ltd, Bury St Edmunds, Suffolk
Printed in England by Clays Ltd, St Ives plc

978–0–141–02326–7

www.greenpenguin.co.uk

Penguin Books is committed to a sustainable future
for our business, our readers and our planet.
The book in your hands is made from paper
certified by the Forest Stewardship Council.

Mixed Sources
Product group from well-managed
forests and other controlled sources
www.fsc.org Cert no. SA-COC-1592
© 1996 Forest Stewardship Council
FSC

To John Archibald Wheeler
who was never afraid to tackle the big questions

Contents

Preface and Acknowledgements

When I was a PhD student at University College London in the 1960s, my supervisor handed me a curious technical paper to read 'as a bit of light relief' from my major project. The paper (which was never actually published in the form in which I read it) was based on a lecture given in the United States by the young English cosmologist and theoretical physicist Brandon Carter. The subject matter was both radical and unusual. The normal work of a theoretical physicist is to investigate an unsolved problem about a natural phenomenon by applying the laws of physics in the form of mathematical equations, and then trying to solve the equations to see how well they describe the real thing. But Carter was addressing an entirely different sort of problem, having to do with the *forms* of the laws themselves. He asked himself the following question: 'Suppose the laws had been a bit different from what they actually are, in this or that respect – what would the consequences be?' Philosophers call this type of investigation counter-factual analysis, and although fiction writers have long been fond of the device (I recently read a novel in which the Nazis defeated Britain in the Second World War and the UK became a German puppet state), it was ground-breaking for a scientist to consider.

The focus of Carter's 'what if' analysis was again unusual for a theoretical physicist. It concerned the existence of life. Specifically, Carter's calculations suggested that if the laws had differed only slightly from what we find them to be, then life would not have been possible and the universe would have gone unobserved. In effect, said Carter, our existence hinges on a certain amount of delicate 'fine tuning' of the laws. Like Goldilocks' porridge, the laws of physics seemed to Carter to be 'just right' for life. It looked like a fix – a big fix. Somewhat unwisely, he named this fine tuning 'the anthropic principle', giving the false impression that it concerned humankind specifically (which was never his intention).

Although Carter's paper was modest in scope and cautious in conclusion, it triggered nothing less than a revolution in scientific thinking, and sparked a furious controversy that has rent the scientific community ever since. The study of counter-factual analysis in physics and cosmology was taken up in the 1970s by Martin Rees and Bernard Carr, resulting in a landmark review paper published in 1979.[1] Inspired by this paper, I wrote a little book on the subject called *The Accidental Universe*, which was published by Cambridge University Press in 1982. A few years later, a much more systematic and thoroughgoing text appeared – *The Anthropic Cosmological Principle*, by John Barrow and Frank Tipler.[2] It has formed the starting point for hundreds of papers over the years.

During the early 1980s, the anthropic principle was slammed by many scientists as quasi-religious mumbo-jumbo. In a scathing put-down in *The New York Review of Books* in 1986, the mathematician and writer Martin Gardner itemized the various proposed versions of the anthropic principle (AP): Weak (WAP), Strong (SAP), Participatory (PAP), Final (FAP) and – his favoured version – the Completely Ridiculous Anthropic Principle (CRAP).[3] And that was pretty much the tone of the debate for a decade or so. But developments in high-energy particle physics and cosmology, especially in the study of the hot big bang that gave birth to the universe, slowly changed sentiment. The laws of physics, once regarded as cast in tablets of stone, began to look less absolute. Evidence accumulated that some of the laws at least were not true, fundamental laws, but 'effective laws', the familiar form of which applies only at energies which are very low compared with the fierce violence of the big bang. Significantly, theoretical analysis suggested that some features of the laws might be accidental, reflecting the vagaries of the manner in which our patch of the universe cooled from the big bang. The implication was, of course, that the low-energy form of these laws could have been different, and might even *be* different, in some other cosmic region. What we had previously been calling 'the universe' began to resemble a variegated 'multiverse' – 'a crazy quilt of environments with different properties and different laws of physics', in the words of Leonard Susskind, a theoretical physicist and cosmologist at Stanford University and a leading proponent of the multiverse idea.[4] It would of course be no surprise that we find ourselves living in a region fit for life, for we obviously could not be living in a place where life is impossible.

At this stage, atheists began to take an interest. Unhappy that the fine tuning of the laws of physics smacked of some sort of divine design, they seized on the multiverse theory as a neat explanation for the uncanny bio-friendliness of the universe. So, confusingly, the anthropic principle came to be seen, at one and the same time, as both a scientific alternative to design and a quasi-religious theory. I stepped into this muddle in 2003, persuading the John Templeton Foundation to sponsor a workshop on multiverse cosmology at Stanford University, which I co-chaired with the cosmologist Andrei Linde. The results of our deliberations were published in a volume edited by Bernard Carr.[5] A follow-up workshop, with more emphasis on string theory (the currently fashionable attempt to unify physics) was held in March 2005.

While these theoretical developments were taking place, some spectacular advances were being made in observational cosmology. These came about from increasingly painstaking surveys of the universe by the Hubble Space Telescope and various ground-based instruments, the detailed mapping of the cosmic afterglow of the big bang by a satellite named WMAP, and the unexpected discovery that the universe is accelerating under the action of some mysterious 'dark energy'. As a result of this fillip, cosmology, long a scientific backwater, suddenly became a mainstream science, with a ferment of new ideas, many of them weird and counter-intuitive. It seems that we are now entering a new era which is transforming our view of the universe and the place of humankind within it.

In this book I shall explain the ideas that underlie these dramatic developments, focusing especially on 'the Goldilocks factor' – the fitness of the universe for life. In the early chapters I shall set out the basic concepts of modern physics and cosmology, and then describe the multiverse theory and the arguments for and against it. Towards the end of the book I shall take a critical look at the various responses to the fine-tuning issue. I shall also ask whether scientists really are on the verge of producing a theory of everything – a complete and self-contained explanation for the entire physical universe – or whether there will always remain a mystery at the heart of existence.

For these later chapters I have drawn inspiration from the great theoretical physicist John Archibald Wheeler, to whom I have dedicated this book. I first learned of Wheeler's work while I was a student, and in subsequent years I came to know him quite well, on both the personal and the professional level. I visited him in Austin, Texas, and he visited me in England on a number of occasions. He graciously endorsed my first book, *The Physics of Time Asymmetry*, with enthusiastic praise, and took a keen interest in my work over three decades. It was a privilege to assist in the organization of his ninetieth-birthday party conference in March 2002, a gathering of extremely distinguished scientists in Princeton, New Jersey, where Wheeler began and ended his career.

In the late 1930s Wheeler worked with the legendary Niels Bohr on key aspects of nuclear fission. He went on to manage the rebirth of gravitational theory in the 1950s, taking up where Einstein left off. It was Wheeler who coined the terms 'black hole' and 'wormhole'. Above all, he recognized the need to reconcile the twin pillars of twentieth-century physics – the general theory of relativity and quantum mechanics – in a unified theory of quantum gravity. Many of his graduate students have enjoyed scientific careers of immense distinction; one of them was the well-known Nobel prize-winner Richard Feynman.

Wheeler's style was distinctive. He was the master of the 'thought experiment', taking an accepted idea and extrapolating it to the ultimate extreme, to see if and when it would break down. He loved to focus on the really big questions: whether physics could be unified; whether space and time could be derived from some more basic entity; whether causality could operate backwards in time; whether the complex and abstract laws of physics could be reduced to a single, simple statement of the obvious; and how observers fitted into the scheme. Not content with simply applying the laws of quantum mechanics, he wanted to know where they came from: 'How come the quantum?'

he asked. Unhappy with the disjunction between the concepts of matter and information, he proposed the idea of 'it from bit' – the emergence of particles from informational bits. Most ambitious of all was his question 'How come existence?' – an attempt to explain *everything* without resorting to some fixed foundation for physical reality that had to be accepted as 'given'.

I once asked Wheeler what he considered his most important achievement, and he answered 'Mutability!' By this he meant that nothing is absolute, nothing is so fundamental that it cannot change under suitably extreme circumstances – and that includes the very laws of the universe. These concepts together led him to propose 'the participatory universe', an idea (or, as Wheeler preferred, 'an idea for an idea') which has proved to be an important part of the multiverse/anthropic discussion. In his beliefs and attitudes, Wheeler represented a large section of the scientific community: committed wholeheartedly to the scientific method of inquiry, but not afraid to tackle deep philosophical questions; not conventionally religious, but inspired by a reverence for nature and a deep sense that human beings are part of a grand scheme which we glimpse only incompletely; bold enough to follow the laws of physics wherever they lead, but not so arrogant as to think that we have all the answers.

I have tried to keep the level of explanation in this book as non-technical as possible, by avoiding jargon and unnecessarily pedantic descriptions. Equations are kept to an absolute minimum. Here and there I have used boxes to summarize or expand some difficult topics. In some ways this book is a sequel to my earlier work *The Mind of God*,[6] but in spite of the emphasis on the deep and meaningful I intend it also to serve as a straightforward introduction to modern cosmology and fundamental physics. I have drawn clear distinctions between secure facts, reasonable theorizing and wild conjecture. The primary purpose of the book is to appeal to scientific inquiry and reason in order to address the big questions of existence. I have made no attempt to consider other modes of discovery, such as mysticism, spiritual enlightenment or revelation through religious experience.

Many people have assisted me in this project. First and foremost was my wife Pauline, who has an uncompromising attitude to sloppy reasoning or unjustified assumptions, and a meticulous attention to detail. She read the text with extraordinary thoroughness, pouncing on many a non sequitur or confusing explanation and chiding me for my irrepressible tendency to lapse into starry-eyed philosophizing. (She also complained that the book stopped just when it was getting interesting.) Having such a hard-headed critic close at hand has improved the book enormously. My literary agent, John Brockman, was the driving force behind the project, having perceived that cosmology is at a crossroads, and the reading public hopelessly confused about the plethora of new discoveries and theories. I have benefited greatly from the participants in the two Stanford workshops, especially Andrei Linde. I am grateful to the John Templeton Foundation for making these lively events possible. Over the years, several people have influenced my thinking, in many cases from personal contact and discussions as well as through their written work. They include John Barrow, Bernard Carr, Brandon Carter, David Deutsch, Michael Duff, George Ellis, David Gross, John Leslie, Charles Lineweaver, Martin Rees, Frank Tipler and, of course, John Wheeler. I should also like to thank Chris Forbes for comments on part of the manuscript and John Woodruff for his meticulous care with the copy-editing.

P. C. W. D.
Sydney, December 2005

A Note on Numbers

In this book I often have to deal with very large and very small numbers. In many cases I write out these numbers in words, but where necessary I use the conventional powers-of-ten notation, as follows:

One million	1,000,000	10^6
One billion	1,000,000,000	10^9
One trillion	1,000,000,000,000	10^{12}
One-millionth	1/1,000,000	10^{-6}
One-billionth	1/1,000,000,000	10^{-9}
One-trillionth	1/1,000,000,000,000	10^{-12}

I

The Big Questions

Confronting the mystery of existence

For thousands of years, human beings have contemplated the world about them and asked the great questions of existence: Why are we here? How did the universe begin? How will it end? How is the world put together? Why is it the way it is? For all of recorded human history, people have sought answers to such 'ultimate' questions in religion and philosophy, or declared them to be completely beyond human comprehension. Today, however, many of these big questions are part of science, and some scientists claim that they may be on the verge of providing answers.

Two major developments have bolstered scientists' confidence that the answers lie within their grasp. The first is the enormous progress made in cosmology – the study of the large-scale structure and evolution of the universe. Observations made using satellites, the Hubble Space Telescope, and sophisticated ground-based instruments have combined to transform our view of the universe and the place of human beings within it. The second development is the growing understanding of the microscopic world within the atom – the subject known as high-energy particle physics. It is mostly carried out with giant particle accelerator machines (what were once called 'atom smashers') of the sort found at Fermilab near Chicago and the CERN Laboratory just outside Geneva. Combining these two subjects – the science of the very large and the science of the very small – provides tantalizing clues that deep and previously unsuspected linkages bind the micro-world to the macro-world. Cosmologists are fond of saying that the big bang, which gave birth to the universe billions of years ago, was the greatest ever particle physics experiment. These spectacular advances hint at a much grander synthesis: nothing less than a complete and unified description of nature, a final 'theory of everything' in which a flawless account of the entire physical world is encompassed within a single explanatory scheme.

The universe is bio-friendly

One of the most significant facts – arguably *the* most significant fact – about the universe is that we are part of it. I should say right at the outset that a great many scientists and philosophers fervently disagree with this statement: that is, they do not think that either life or consciousness is even remotely significant in the great cosmic scheme of things. My position, however, is that I take life and mind (i.e. consciousness) seriously, for reasons I shall explain in due course. At first sight life seems to be irrelevant to the subject of cosmology. To be sure, the surface of the Earth has been modified by life, but in the grand sweep of the cosmos our planet is but an infinitesimal dot. There is an indirect sense, however, in which the existence of life in the universe *is* an important cosmological fact. For life to emerge, and then to evolve into conscious beings like ourselves, certain conditions have to be satisfied. Among the many prerequisites for life – at least, for life as we know it – is a good supply of the various chemical elements needed to make biomass. Carbon is the key life-giving element, but oxygen, hydrogen, nitrogen, sulphur and phosphorus are crucial too. Liquid water is another essential ingredient. Life also requires an energy source, and a stable environment, which in our case are provided by the sun. For life to evolve past the level of simple microbes, this life-encouraging setting has to remain benign for a very long time; it took billions of years for life on Earth to reach the point of intelligence.

On a larger scale, the universe must be sufficiently old and cool to permit complex chemistry. It has to be orderly enough to allow the untrammelled formation of galaxies and stars. There have to be the right sorts of forces acting between particles of matter to make stable atoms, complex molecules, planets and stars. If almost any of the basic features of the universe, from the properties of atoms to the distribution of the galaxies, were different, life would very probably be impossible.[1] Now, it happens that to meet these various requirements, certain stringent conditions must be satisfied in the underlying laws of physics that regulate the universe, so stringent in fact that a bio-friendly universe looks like a fix – or 'a put-up job', to use the pithy description of the late British cosmologist Fred Hoyle. It appeared to Hoyle as if a super-intellect had been 'monkeying' with the laws of physics.[2] He was right in his impression. On the face of it, the universe *does* look as if it has been designed by an intelligent creator expressly for the purpose of spawning sentient beings. Like the porridge in the tale of Goldilocks and the three bears, the universe seems to be 'just right' for life, in many intriguing ways. No scientific explanation for the universe can be deemed complete unless it accounts for this appearance of judicious design.

Until recently, 'the Goldilocks factor' was almost completely ignored by scientists. Now, that is changing fast. As I shall discuss in the following chapters, science is at last coming to grips with the enigma of why the universe is so uncannily fit for life. The explanation entails understanding how the universe began and evolved into its present form, and knowing what matter is made of and how it is shaped and structured by the different forces of nature. Above all, it requires us to probe the very nature of physical laws.

The cosmic code

Throughout history, prominent thinkers have been convinced that the everyday world observed through our senses represents only the surface manifestation of a deeper hidden reality, where the answers to the great questions of existence should be sought. So compelling has been this belief that entire societies have been shaped by it. Truth-seekers have practised complex rituals and rites, used drugs and meditation to enter trance-like states, and consulted shamans, mystics and priests in an attempt to lift the veil on a shadowy world that lies beneath the one we perceive. The word 'occult' originally meant 'knowledge of concealed truth', and seeking a gateway to the occult domain has been a major preoccupation of all cultures, ranging from the Dreaming of Aboriginal Australians to the myth of Adam and Eve tasting the forbidden fruit of the tree of knowledge.

The advent of reasoned argument and logic did nothing to dispel the beguiling notion of a hidden reality. The ancient Greek philosopher Plato compared the world of appearances to a shadow playing on the wall of a cave. Followers of Pythagoras were convinced that numbers possess mystical significance. The Bible is also replete with numerology, for example the frequent appearances of 7 and 40, or the association of 666 with Satan. The power of numbers led to a belief that certain integers, geometrical shapes and formulas could invoke contact with a supernatural plane, and that obscure codes known only to initiates might unlock momentous cosmic secrets.[3] Remnants of ancient numerology survive today: some superstitious people still believe that numbers such as 8 and 13 are lucky or unlucky.

Attempts to gain useful information about the world through magic, mysticism and secret mathematical codes mostly led nowhere. But about three hundred and fifty years ago, the greatest magician who ever lived finally stumbled on the key to the universe – a cosmic code that would open the floodgates of knowledge. This was Isaac Newton – mystic, theologian and alchemist – and in spite of his mystical leanings, he did more than anyone to change the age of magic into the age of science. Newton, together with a small number of other scientific luminaries who included Nicolaus Copernicus, Johannes Kepler and Galileo Galilei, gave birth to the modern scientific age. The word 'science' is derived from the Latin *scientia*, simply meaning 'knowledge'. Originally it was just one of many arcane methods that were used to probe beyond the limitations of our senses in the hope of accessing an unseen reality. The particular brand of 'magic' employed by the early scientists involved hitherto unfamiliar and specialized procedures, such as manipulating mathematical symbols on pieces of paper and coaxing matter to behave in strange ways. Today we take such practices for granted, and call them scientific theory and experiment. No longer is the scientific method of inquiry regarded as a branch of magic, the obscure dabbling of a closed and privileged priesthood. But familiarity breeds contempt, and these days the significance of the scientific process is often under-appreciated. In particular, people show little surprise that science actually *works*, and that we really are in possession of the key to the universe. The ancients were right: beneath the surface complexity of nature lies a hidden subtext, written in a subtle mathematical code. This cosmic code[4] contains the secret rules on which the universe runs. Newton, Galileo and other early scientists treated their investigations as a religious quest. They thought that by exposing the patterns woven into the processes of nature they truly were glimpsing the mind of God.[5] Modern scientists are mostly not religious, yet they still accept that an intelligible script underlies the workings of nature, for to believe otherwise would undermine the very motivation for doing research, which is to uncover something meaningful about the world that we don't already know.

Finding the key to the universe was by no means inevitable. For a start, there is no logical reason why nature should have a mathematical subtext in the first place. And even if it does, there is no obvious reason why humans should be capable of comprehending it. You would never guess by looking at the physical world that beneath the surface hubbub of natural phenomena lies an abstract order, an order that can't be seen or heard or felt, but *deduced*. Even the wisest mind couldn't tell merely from daily experience that the diverse physical systems making up the cosmos are linked, deep down, by a network of coded mathematical relationships. Yet science has uncovered the existence of this concealed mathematical domain. We human beings have been made privy to the deepest workings of the universe. Other animals observe the same natural phenomena as we do, but alone among the creatures on this planet, *Homo sapiens* can also *explain* them.

How has this come about? Somehow the universe has engineered, not just its own awareness, but its own *comprehension*. Mindless, blundering atoms have conspired to make, not just life, not just mind, but *understanding*. The evolving cosmos has spawned beings who are able not merely to watch the show, but to unravel the plot. What is it that enables something as small and delicate and adapted to terrestrial life as the human brain to engage with the totality of the cosmos and the silent mathematical tune to which it dances? For all we know, this is the first and only time anywhere in the universe that minds have glimpsed the cosmic code. If humans are snuffed out in the twinkling of a cosmic eye, it may never happen again. The universe may endure for a trillion years, shrouded in total mystery, save for a fleeting pulse of enlightenment on one small planet around one average star in one unexceptional galaxy, 13.7 billion years after it all began.

Could it just be a fluke? Might the fact that the deepest level of reality has connected to a quirky natural phenomenon we call 'the human mind' represent nothing but a bizarre and temporary aberration in an absurd and pointless universe? Or is there an even deeper sub-plot at work?

The concept of laws

I may have given the impression that Newton belonged to a small sect that conjured science out of the blue as a result of mystical investigation. This wasn't so. Their work did not take place in a cultural vacuum: it was the product of many ancient traditions. One of these was Greek philosophy, which encouraged the belief that the world could be explained by logic, reasoning and mathematics. Another was agriculture, from which people learned about order and chaos by observing the cycles and rhythms of nature, punctuated by sudden and unpredictable disasters. And then there were religions, especially monotheistic faiths, which encouraged belief in a created world order. The founding assumption of science is that the physical universe is neither arbitrary nor absurd; it is not just a meaningless jumble of objects and phenomena haphazardly juxtaposed. Rather, there is a coherent *scheme of things*. This is often expressed by the simple aphorism that there is order in nature. But scientists have gone beyond this vague notion to formulate a system of well-defined *laws*.

The existence of laws of nature is the starting point of this book, and indeed it is the starting point of science itself. But right at the outset we encounter an obvious and profound enigma:

Where do the laws of nature come from?

As I have remarked, Galileo, Newton and their contemporaries regarded the laws as thoughts in the mind of God, and their elegant mathematical form as a manifestation of God's rational plan for the universe. Few scientists today would describe the laws of nature using such quaint language. Yet the questions remain of what these laws are and why they have the form that they do. If they aren't the product of divine providence, how can they be explained?

Historically, laws of nature were discussed by analogy to civil law, which arose as a means of regulating human society. Civil law is a concept that dates back to the time of the first settled communities, when some form of authority was needed to prevent social disorder. Typically, a despotic leader would concoct a set of rules and exhort the populace to comply with them. Since one person's rules can be another person's problem, rulers would often appeal to divine authority to buttress their power. A city's god might be literally a stone statue in the town square, and a priest would be appointed to interpret the god's commandments. The notion of turning to a higher, non-material authority as justification for civil law underpins the Ten Commandments and was refined in the Jewish Torah. Remnants of this notion survived into the modern era as the concept of the Divine Right of Kings.

Appeal was also made to an invisible higher power in support of laws of nature. In the fourth century BCE the Stoic philosopher Cleanthes described 'Universal Nature, piloting all things according to Law'.[6] The order of nature was perhaps clearest in the heavens – the very domain of the gods. Indeed, the word 'astronomy' means 'law of the stars'. The first-century BCE Roman poet Lucretius referred to the way in which nature requires 'each thing to abide by the law that governs its creation'.[7] In the first century CE, Marcus Manilius was explicit about the source of nature's order, writing that 'God brought the whole universe under law'.[8] It was a position wholeheartedly embraced by the monotheistic religions: God the Creator was also God the Lawmaker, who ordered nature according to his divine purposes. Thus the early Christian theologian Augustine of Hippo wrote that 'the ordinary course of nature in the whole of creation has certain natural laws'.[9]

By the thirteenth century, European theologians and scholars such as Roger Bacon had arrived at the conclusion that laws of nature possess a mathematical basis, a notion that dates back to the Pythagoreans. Oxford University became the centre for scholars who applied mathematical philosophy to the study of nature. One of these so-called Oxford Calculators was Thomas Bradwardine (1295–1349), later to become Archbishop of Canterbury. Bradwardine has been credited with the first scientific work to announce a general mathematical law of physics in the modern sense. Given this background, it is no surprise that when modern science emerged in Christian Europe in the sixteenth and seventeenth centuries, it was perfectly natural for the early scientists to believe that the laws they were discovering in the heavens and on Earth were the mathematical manifestations of God's ingenious handiwork.

The special status of the laws of physics

Today, the laws of *physics* occupy the central position in science; indeed, they have assumed an almost deistic status themselves, often cited as the bedrock of physical reality. Let me give an everyday example. If you go to Pisa in Italy, you can see the famous leaning tower (now restored to a safe inclination by engineering works). Tradition says that Galileo dropped balls from the top of the tower to demonstrate how they fall under gravity. Whether or not this is true, he certainly did carry out some careful experiments with falling bodies, which is how he came to discover the following law. If you drop a ball from the top of a tall building and measure how far it falls in one second, then repeat the experiment for two seconds, three seconds, and so on, you will find that the distance the ball travels increases as the *square* of the time. The ball will fall four times as far in two seconds as in one, nine times as far in three seconds, and so on. Schoolchildren learn about this law as 'a fact of nature', and normally move on without giving it much further thought. But I want to stop right there and ask the question, *why?* Why is there such a mathematical rule at work on falling bodies? Where does the rule come from? And why that rule and not some other?

Let me give another example of a law of physics, one that made a big impression on me in my schooldays. It concerns the way magnets lose their grip on each other with separation. Line them up side by side and measure the force as the distance between them increases. You will find that the force diminishes with the *cube* of the distance, which is to say that if we double the distance between the magnets, the force falls to one-eighth, treble it and the force will be $1/27$, and so on. Again, I am prompted to ask the question, *why?*

Some laws of physics bear the name of their discoverer, such as Boyle's law for gases, which tells you that if you double the volume of a fixed mass of gas while keeping the temperature constant, its pressure is halved. Or Kepler's laws of planetary motion, one of which says that the square of the period of an orbit is proportional to the cube of the orbit's radius. Perhaps the best-known laws are Newton's laws of motion and gravitation, the latter supposedly inspired by an apple falling from a tree. It states that the force of gravity diminishes with distance as the *square* of the separation between the two bodies. That is, the force that binds the Earth to the sun, and prevents it from flying off alone across the galaxy, would fall to only one-quarter the strength if the Earth's orbit were twice as big. This is known as an 'inverse square law'. I have drawn a graph depicting it in Figure 1.

The fact that the physical world conforms to mathematical laws led Galileo to make a famous remark. 'The great book of nature', he wrote, 'can be read only by those who know the language in which it was written. And this language is mathematics.'[10] The same point was made more bluntly three centuries later by the English astronomer James Jeans: 'The universe appears to have been designed by a pure mathematician.'[11] It is the mathematical aspect that makes possible what physicists mean by the much-misunderstood word 'theory'. Theoretical physics entails writing down equations that capture (or model, as scientists say) the real world of experience in a mathematical world of numbers and algebraic formulas. Then, by manipulating the mathematical symbols, one can work out what will happen in the real world, without actually carrying out the observation. That is, by applying the equations that express the laws relevant to the problem of interest, the theoretical physicist can predict the answer. For example, by using Newton's laws of motion and gravitation, engineers

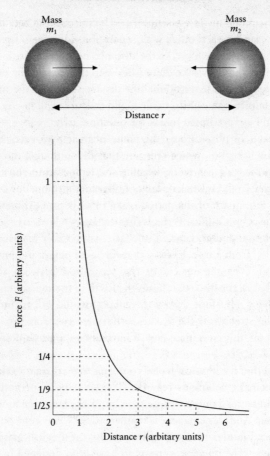

1. *Inverse-square law of gravity*

The gravitational force between two masses m_1 and m_2 (they might be stars or planets) diminishes with the distance between their centres of mass according to the simple curve shown.

can figure out when a spacecraft launched from Earth will reach Mars. They can also calculate the required mass of fuel, the most favourable orbit and a host of other factors, in advance of the mission. And it works! The mathematical model faithfully describes what actually happens in the real world. (Of course, in practice one may have to simplify the model to save time and cost of the analysis, making the predictions good only to a certain level of approximation, but that is not the fault of the laws.)

When I was at school I took a fancy to a young lady in my class named Lindsay. I didn't see much of her because she was studying mainly arts subjects and I was studying the sciences and mathematics. But we did meet up in the school library from time to time. On one occasion I was busy doing a calculation. I even remember what it was. If you throw a ball in the air at a certain speed and angle, Newton's laws let you work out how far it will travel before it hits the ground. The equations tell you that to achieve maximum range you should throw the ball at 45° to the horizontal. If the ground on which you are standing slopes upwards, however, the angle needs to be greater; by how much depends on the amount of slope. I was deeply engrossed calculating the maximum range up an inclined plane when Lindsay looked up and asked what I was doing. I explained. She seemed puzzled and sceptical. 'How can you possibly know what a ball will do by writing things on a sheet of paper?' she asked. At the time I dismissed her question as silly – after all, this was what we had been taught to do! But over the years I came to see that her impulsive response precisely captures one of the deepest mysteries of science: *Why* is nature shadowed by a mathematical reality? Why does theoretical physics work?[12]

How many laws are there?

As scientists have probed deeper and deeper into the workings of nature, all sorts of laws have come to light that are not at all obvious from a casual inspection of the world, for example laws that regulate the internal components of atoms or the structure of stars. The multiplicity of laws raises another challenging question: How long would a complete list of laws be? Ten? Twenty? Two hundred? Might the list even be infinitely long?

Not all the laws are independent of one another. It wasn't long after Galileo, Kepler, Newton and Boyle began discovering laws of physics that scientists found links between them. For example,

Newton's laws of gravitation and motion *explain* Kepler's three laws of planetary motion, and so are in some sense deeper and more powerful. Newton's laws of motion also explain Boyle's law of gases when they are applied in a statistical way to a large collection of chaotically moving molecules.

In the four centuries that have passed since the first laws of physics were discovered, more and more have come to light, but more and more links have been spotted too. The laws of electricity, for example, were found to be connected to the laws of magnetism, which in turn explained the laws of light. These interconnections led to a certain amount of confusion about which laws were 'primary' and which could be derived from others. Physicists began talking about 'fundamental' laws and 'secondary' laws, with the implication that the latter were formulated for convenience only. Sometimes physicists call these 'effective laws' to distinguish them from the 'true', underlying fundamental laws, to which, at least in principle, the effective, or secondary, laws can all be subsumed. In this respect, the laws of physics differ markedly from the laws of civil society, which are an untidy hotchpotch of statutes expanding without limit. To take an extreme case, the tax laws in most countries run to millions of words of text. By comparison, the Great Rule Book of Nature (at least as it is currently understood) would fit comfortably onto a single page. This streamlining and repackaging process – finding links between laws, and reducing them to ever more fundamental laws – continues apace, and it's tempting to believe that, at rock bottom, there is just a handful of *truly* fundamental laws, possibly even a single super-law, from which all the other laws derive.

Given that the laws of physics underpin the entire scientific enterprise, it is curious that very few scientists bother to ask what these laws actually mean. Speak to physicists, and most of them will talk as if the laws are *real things* – not physical objects of course, but abstract relationships between physical entities. Importantly, though, they are relationships that really exist, 'out there' in the world, and not just in our heads.

For brevity I have been a bit cavalier with my terminology. If you confront a physicist and say, 'Show me the laws of physics,' you will be referred to a collection of textbooks – on mechanics, gravitation, electromagnetism, nuclear physics, and so on. But a pertinent question is whether the laws you find in the books are actually *the* laws of physics, or just somebody's best stab at them. Few physicists would claim that a law found in a book which is in print today is the last word on the subject; all the textbook laws are probably just some sort of approximation to the real ones. Most physicists nevertheless believe that as science advances, so the textbook laws will converge on the Real Thing.[13]

Are the laws real?

There is a subtlety buried in all this which will turn out to be of paramount importance when I come to discuss the origin of the laws. The idea of laws began as a way of formalizing patterns in nature that connect together physical events. Physicists became so familiar with the laws that somewhere along the way the laws themselves – as opposed to the events they describe – became promoted to reality. The laws took on a life of their own. It is hard for non-scientists to grasp the significance of this step. One analogy might be with the world of finance. Money in the pocket means coins and notes – real physical things that get exchanged for real physical goods or services. But money in the abstract has also taken on a life of its own. Investors can grow (or shrink, in my case) money without ever buying or selling physical stuff. For example, there are rules for manipulating different currencies that are at best tenuously connected to the actual purchasing function in your local corner shop. In fact, there is far more 'money' in circulation, much of it swirling around cyberspace via the internet, than can ever be accumulated as coins and notes. In a similar vein, the laws of physics inhabit an abstract realm and touch the physical world only when they 'act'. It's almost as if the laws are lying in wait, ready to seize control of a physical process and compel it to comply, just as the rules of monetary conversion are 'in place' even when nobody is actually converting anything. This 'prescriptive' view of physical laws as having power over nature is not without its detractors (namely, philosophers who prefer a 'descriptive' view).[14] But most physicists working on fundamental topics inhabit the prescriptive camp, even if they won't own up to it explicitly.

So we have this image of really-existing laws of physics ensconced in a transcendent eyrie, lording it over lowly matter. One reason for this way of thinking about the laws concerns the role of mathematics. Numbers began as a way of labelling and tallying physical things such as beads or sheep. As the subject of mathematics developed, and extended from simple arithmetic into geometry, algebra, calculus, and so forth, so these mathematical objects and relationships came to assume an independent existence. Mathematicians believe that statements such as '$3 \times 5 = 15$' and '11 is a prime number' are inherently true – in some absolute and general sense – without being restricted to 'three sheep' or 'eleven beads'.

Plato considered the status of mathematical objects, and chose to locate numbers and idealized geometrical shapes in an abstract realm of perfect forms. In this Platonic heaven there would be found, for example, perfect circles – as opposed to the circles we encounter in the real world, which will always be flawed approximations to the ideal. Many modern mathematicians are Platonists (at least at weekends). They believe that mathematical objects have real existence, yet are not situated in the physical universe. Theoretical physicists, who are steeped in the Platonic tradition, also find it natural to locate the mathematical laws of physics in a Platonic realm. I have depicted this arrangement diagrammatically in Figure 2. In the final chapter I shall take a critical look at the nature of physical laws and ask whether the Platonic view has become an unwelcome fixation in the drive to understand the mathematical underpinnings of the universe.

Goodbye God?

Religion was the first systematic attempt to explain the universe comprehensively. It presented the world as a product of mind or minds, of supernatural agents who could order or disorder nature at will. In Hinduism, Brahma is creator and Shiva destroyer. In Judaism, Yahweh is both creator and destroyer. For the traditional Aboriginal

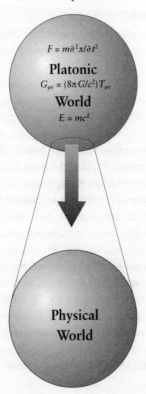

2. Where are the laws of physics located?

Plato believed that mathematical objects really exist, located not in the physical world of appearances but in an abstract realm of idealized forms, accessible to the intellect. Theoretical physicists, who express the laws of physics as mathematical equations, tend to follow this tradition. They prefer to envisage the laws of physics as having real existence, but transcending *physical* reality.

people of the Kimberley in Australia, two creator beings acted in synergy. Wallanganda, a male space being, sprinkled water on Wunngud, a female snake coiled in jelly, to make Yorro Yorro – the world as we see it.[15] In these sorts of schemes, things are as they are because a god (or gods) decided they should be so. The major world religions devoted centuries of scholarship in attempts to make these theistic explanations cogent and consistent. Even today, millions of people base their world view on a religious interpretation of nature.

Science was the second great attempt to explain the world. This time, explanations were cast in terms of impersonal forces and natural, physical processes rather than the activities of purposive supernatural agents. When scientific explanations conflicted with religious explanations, religion invariably lost the battle. Mostly, theologians retreated to concentrate on social and ethical matters such as spiritual enlightenment, content to leave interpreting the physical universe to the scientists. There are still people who believe that rain is made by rain gods rather than by atmospheric processes, but I wouldn't rate their chances in a debate with a meteorologist. When it comes to actual physical phenomena, science wins hands down against gods and miracles. That is not to say that science has explained everything. There remain some pretty big gaps: for example, scientists don't know how life began, and they are almost totally baffled by consciousness. Even some familiar phenomena, such as turbulent fluids, are not completely understood. But this doesn't mean that one needs to appeal to magic or miracles to plug the gaps; what is needed are advances in scientific understanding. This is a topic I shall address in detail in Chapter 10.

When it comes to metaphysical questions such as 'Why are there laws of nature?' the situation is less clear. These sorts of questions are not much affected by specific scientific discoveries: many of the really big questions have remained unchanged since the birth of civilization, and still vex us today. The various faith traditions have had hundreds of years to ponder them carefully. Religious scholars such as Anselm and Thomas Aquinas were not pious simpletons, but the intellectual heavyweights of their age.

Many scientists who are struggling to construct a fully comprehensive theory of the physical universe openly admit that part of the motivation is to finally get rid of God, whom they view as a dangerous and infantile delusion. And not only God, but any vestige of God-talk, such as 'meaning' or 'purpose' or 'design' in nature. These scientists see religion as so fraudulent and sinister that nothing less than total theological cleansing will do. They concede no middle ground, and regard science and religion as two implacably opposed world views. Victory is assumed to be the inevitable outcome of science's intellectual ascendancy and powerful methodology.

But will God go quietly? Even within the world of organized religion, the concept of 'God' means many different things to different people. At the level of popular, Sunday-school Christianity, God is portrayed simplistically as a sort of Cosmic Magician, conjuring the world into being from nothing and from time to time working miracles to fix problems. Such a being is obviously in flagrant contradiction to the scientific view of the world. The God of scholarly theology, by contrast, is cast in the role of a wise Cosmic Architect whose existence is manifested through the rational order of the cosmos, an order that is in fact *revealed* by science. That sort of God is largely immune from scientific attack.

Is the universe pointless?

Even atheistic scientists will wax lyrical about the scale, the majesty, the harmony, the elegance, the sheer ingenuity of the universe of which they form so small and fragile a part. As the great cosmic drama unfolds before us, it begins to look as though there is a 'script' – a scheme of things – which its evolution is following. We are then bound to ask, who or what wrote the script? Or did the script somehow, miraculously, write itself? Is the great cosmic text laid down once and for all, or is the universe, or the invisible author, making it up as it goes along? Is this the only drama being staged, or is our universe just one of many shows in town?

The fact that the universe conforms to an orderly scheme, and is not an arbitrary muddle of events, prompts one to wonder – God or no God – whether there is some sort of meaning or purpose behind it all. Many scientists are quick to pour scorn even on this weaker suggestion, however. Richard Feynman, arguably the finest theoretical physicist of the mid-twentieth century, thought that 'the great accumulation of understanding as to how the physical world behaves only convinces one that this behavior has a kind of meaninglessness about it'.[16] This sentiment is echoed by the theoretical physicist and cosmologist Steven Weinberg: 'The more the universe seems comprehensible the more it also seems pointless.'[17] Weinberg came in for some flak from his colleagues for writing this comment – not because he denied that the universe had a point, but for even suggesting that it *could* have a point.

To be sure, concepts like 'meaning' and 'purpose' are categories devised by humans, and we must take care when attempting to project them on to the physical universe. But *all* attempts to describe the universe scientifically draw on human concepts: science proceeds precisely by taking concepts that humans have thought up, often from everyday experience, and applying them to nature. Doing science means figuring out what is going on in the world – what the universe is 'up to', what it is 'about'. If it isn't 'about' anything, there would be no good reason to embark on the scientific quest in the first place, because we would have no rational basis for believing that we could thereby uncover additional coherent and meaningful facts about the world. So we might justifiably invert Weinberg's dictum and say that the more the universe seems pointless, the more it also seems incomprehensible. Of course, scientists might be deluded in their belief that they are finding systematic and coherent truth in the workings of nature. Ultimately there may be no reason at all for why things are the way they are. But that would make the universe a fiendishly clever bit of trickery. Can a truly absurd universe so convincingly mimic a meaningful one? This is the biggest of the big questions of existence that we will confront as we embark on our investigation of life, the universe and everything.

Key points

- Many big questions of existence are now on the scientific agenda.
- A really big question is why the universe is fit for life; it looks like it has been 'fixed up'.
- The universe obeys mathematical laws; they are like a hidden subtext in nature. To appreciate this book you have to be comfortable with that idea.

- The mathematical laws of physics underlie everything. Many physicists think they are real, and inhabit a transcendent Platonic realm.
- Science reveals that there is a coherent scheme of things, but scientists do not necessarily interpret that as evidence for meaning or purpose in the universe. Most, but by no means all, scientists are atheists or agnostics.
- Somehow I am supposed to explain all this.

2

The Universe Explained

The big bang and the expanding universe

According to Cambridge University folklore, the nuclear physicist Ernest Rutherford is said to have issued an edict to his subordinates against fanciful and grandiose speculation. 'Don't let me catch anyone talking about the universe in my department!' he warned. That was in the 1930s, and to be fair to Rutherford, cosmology didn't then exist as a proper science. Even when I was a student in London in the 1960s, cynics quipped that there is speculation, speculation squared – and cosmology. Since then, however, advances in telescope design, data processing and the use of satellites have transformed cosmology out of all recognition. This period of rapid progress culminated on 11 February 2003, when the world's press carried a curious-looking oval-shaped picture resembling a Jackson Pollock drip painting. This unprepossessing image provided a window on the universe like nothing available before, and it propelled cosmology – the study of the origin, evolution and fate of the universe on the largest scales of space and time – into the twenty-first century. At last, the subject had matured into a proper, quantitative science. The picture summarized the initial results from a satellite named the Wilkinson Microwave Anisotropy Probe, or WMAP for short. The job of WMAP is to map the sky, using not light but heat: it is a thermal map of the universe, compiled in unprecedented detail. The features imprinted in the image are remnants of the birth of the universe, over 13 billion years ago (see Figure 3).

Cosmology would not exist as a subject unless there were such a

3. *Afterglow of the creation*

This thermal map of the sky at microwave frequencies as measured by the WMAP satellite provides a snapshot of the universe at about 380,000 years after the big bang. The blobs and speckles represent slight temperature variations imprinted in the radiation by density fluctuations in the early universe. By studying the details of these variations, cosmologists can work out a great deal about the origin, history and likely fate of the universe, as well as its make-up and geometry. The over-dense regions, represented by the lighter patches, are the 'seeds' around which clusters of galaxies formed. The image shown here is based on more refined data released in March 2006. (Courtesy NASA/WMAP Science Team.)

thing as 'the universe' to explain. Instead of finding that space is filled with a dog's breakfast of unrelated bric-a-brac, astronomers see an orchestrated and coherent unity. On the largest scale of size there is order and uniformity. Stars and galaxies billions of light years away closely resemble those in our astronomical backyard and are distributed in much the same way everywhere. Their compositions and motions are similar. The laws of physics appear to be identical as far out in space as our instruments can penetrate. In short, there is cosmos rather than chaos. This basic fact is crucial for our existence: life could not emerge, still less evolve to the point of intelligence, amid chaos. It is also – or at least it was until recently – deeply mysterious. Why should the totality of things be organized so systematically? To find the answer to this intriguing question, we need to understand how the universe began and work out how it evolved over billions of years to attain its present orderly and life-encouraging form.

When our ancestors contemplated the heavens they envisaged the sun, moon and stars moving around the Earth. The scale of the cosmos was unknown. Even the invention of the telescope failed to reveal the true magnitude of the universe, and only in the last few decades have astronomers been able to pencil in the numbers that calibrate the vast scale of things. Our sun is one among hundreds of billions of stars that make up the Milky Way galaxy, and the Milky Way is in turn just one among hundreds of billions of galaxies scattered through space to the limits of our instruments. The gaps between stars are so large that astronomers measure them in light years – the distance light travels in one year. One light year works out to be about 6 trillion miles or 10 trillion kilometres. To put this into perspective, the moon is just over a light second away, and the sun a little more than 8 light minutes. The Milky Way, which is a typical spiral galaxy, measures about 100,000 light years across. The Andromeda Galaxy, a near neighbour of the Milky Way, lies at a distance of about $2^{1}/_{2}$ million light years. The farthest galaxies imaged in the Hubble Space Telescope are over 10 billion light years away. In human terms, the universe is almost unimaginably vast.

Although the results from WMAP mark the moment that observational cosmology came of age, the birth of the subject goes back 80 years, to the ground-breaking work of the lawyer-turned-astronomer Edwin Hubble. It is he who is credited with the discovery that the universe is expanding, although many of the crucial observations were carried out by the astronomer Vesto Slipher. Hubble and Slipher studied light from many galaxies and found that the more distant ones were redder. It had long been known that light waves from a receding source would be stretched, and therefore shifted towards the red end of the spectrum (conversely, light from an approaching source is blue-shifted). Hubble and Slipher found that the red shift gets bigger the farther away from us a galaxy is located and that, furthermore, the effect is the same in all directions. The simplest explanation of these facts, and the one that Hubble announced to the world, is that the galaxies are rushing away from us in an orderly pattern of expansion.

Obviously, if the universe is expanding now it must have been more compressed in the past. Using the measured rate of expansion, it is easy enough to run the great cosmic movie backwards (as a theoretical exercise!) to determine that, many billions of years ago, all the galaxies would have been squeezed into one place. This suggests that the universe, at least in the form we know, began with a vast explosion from a highly dense state, an event now known as the big bang. Curiously, this appellation was originally a term of derision introduced in the 1950s by Fred Hoyle, who never accepted the theory. Today, with accurate observations by instruments such as WMAP and the Hubble Space Telescope, we can put quite a precise date on the big bang. The best current estimate is 13.7 billion years ago. For comparison, the Earth is 4.56 billion years old.

The afterglow of creation

If the universe was once very dense, it should also have been very hot, since matter heats up when compressed and cools when expanded. But hot matter emits thermal radiation (think of the radiant heat from the sun, or from the glowing embers of a fire), so we can expect the heat left over from the birth of the universe to be bathing the universe today in a faint glow of radiation. And indeed it is. In 1967, two radio engineers working on satellite communications for Bell Laboratories in the United States, Arno Penzias and Robert Wilson, stumbled across radiation coming from space that was soon identified as the expected relic of the big bang. That discovery marked the tipping point when scientists finally sat up and took notice of the big bang theory. The radiation is evenly distributed across the sky at a temperature[1] of 2.725 K, or about *minus* 270°C – so don't expect the sky to be glowing dull red. Radiation at this temperature lies mainly in the microwave region of the electromagnetic spectrum, so the heat from the big bang is known as the 'cosmic microwave background', usually abbreviated to CMB.

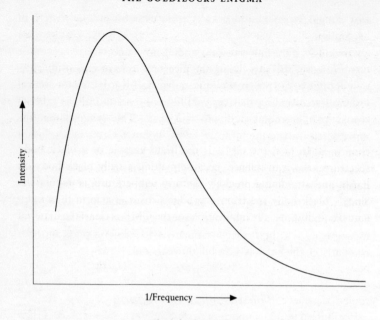

Intensity →

1/Frequency ⟶

4. Black body spectrum

The curve, based on the measurements made by the WMAP satellite, shows how the heat energy left over from the big bang is distributed across a range of wavelengths. The shape of the curve is distinctive, corresponding precisely to the spectrum of radiation from a system at a uniform temperature. It implies that the cosmic microwave background originated from a state of thermodynamic equilibrium in the far past. The observations fit the theory so precisely that the measurement errors are smaller than the thickness of the line in the figure. (Courtesy NASA.)

Although it is hard to imagine what other than a cosmic explosion could have produced the CMB, the real clincher for the big bang theory came when astronomers were able to measure the precise spectrum of the radiation (see Figure 4). If you look at a glowing object, such as a flame or a star, it will emit energy across a range of colours, or wavelengths. Plot the distribution on a graph and you get what is called a spectrum. The spectrum of light or heat emanating from the sun, or a candle flame, is very complicated, with many peaks and troughs in the graph. But the spectrum of a particular type of heat radiation has a simple, distinctive, universal form: the one from an oven in which the interior has reached a precisely uniform temperature. Physicists call this a black body spectrum, because a completely unreflecting body (which would look black at low temperature) will radiate heat with this particular distribution of energy across its different wavelengths. Significantly, the CMB has just such a black body spectrum, as observations by WMAP and other instruments have confirmed. In fact, the CMB is the best example of a black body spectrum known to science, because making a truly black body on Earth, and attaining a precisely uniform temperature, is impossible. Since a black body spectrum is produced by a system in thermodynamic equilibrium,[2] the implication for the CMB is clear: the material of the early universe must have been smoothly distributed through space at the same density and temperature everywhere. The black body spectrum was the smoking gun, confirming that the universe began in a hot, dense, uniform state, from which it expanded and cooled to achieve its present form.

I've glossed over an important part of the story. As our planet orbits the sun, the sun whirls around the galaxy and the galaxy meanders among its neighbours, the Earth finds itself sweeping through the CMB at about 600 kilometres per second. Because of this relative movement, the sky looks a bit hotter in the direction in which we happen to be moving than does the opposite side of the sky. When this effect is subtracted out, however, the radiation is astonishingly smooth. To roughly one part in 100,000, there is no variation across the sky.

Cosmologists knew all along, however, that the CMB couldn't be *precisely* uniform because the universe isn't precisely uniform. Matter is aggregated into galaxies, and galaxies are in turn arranged in clusters and superclusters. Maps of the distribution of galaxies, painstakingly constructed by astronomers using optical telescopes over the past twenty years, reveal clumping on all scales of size except the very largest. Averaged over a billion light years, the universe looks the same everywhere, but on a scale of a hundred million light years or less it's a different story, with aggregations of galaxies standing out prominently. Had the early universe consisted of perfectly uniform gas, there could be no such structure. But over time, even the smallest irregularities in the primordial gases would have become amplified under the action of gravitation. Any region of the universe that was slightly over-dense initially would have drawn in material at the expense of its surroundings, thereby enhancing the density contrast and accelerating the process. The slow implosion of the primordial gases into clumps would eventually have turned into catastrophic collapse were it not for the fact that the universe is also expanding, which serves to dilute the gas and counter the tendency for aggregation. Calculations of these competing effects indicate that to grow galaxies distributed in the observed manner, the universe must have started out with density variations of about one part in 100,000. Because denser gas is more compressed it is hotter, so irregularities in density translate into irregularities in temperature. So the early universe should have possessed tiny temperature variations, if the big bang theory is to hang together consistently. And that is precisely what WMAP found.

So the story goes something like this. The universe began 13.7 billion years ago with a big bang. The state of the early universe was one of extremely hot and dense, ionized,[3] opaque, expanding gas suffused with heat radiation. The gas was distributed through space with almost but not quite perfect uniformity. By about 380,000 years after the big bang, the universe had cooled to a few thousand degrees, and at this point the gas de-ionized (i.e. the nuclei and electrons combined into atoms) as a result of which it became transparent. The heat radiation thereafter was largely unaffected by its passage through matter, and it has travelled almost freely ever since.[4] Therefore, when astronomers detect the CMB they are glimpsing the universe as it was about 380,000 years after the big bang. In effect, the CMB is a snapshot of what the universe was like when it was less than 0.003 per cent of its present age. The tiny variations in temperature detected by WMAP represent the seeds of cosmic structure without which there would have been no galaxies, stars, planets – or astronomers. So this is another one of those 'convenient' facts that makes the universe bio-friendly, and which needs explaining.

Where is the centre of the universe?

Popular accounts of the big bang often describe it as the detonation of a compact ball of matter poised in a pre-existing void, with the galaxies compared to fragments flying away from the centre of the explosion. Easy though this image may be to grasp, it is seriously misleading, and the source of much confusion: people are inevitably prompted to ask, 'Where is the centre of the universe?' The following e-mail message I recently received is fairly typical:

Has the proof of expanding universe been realized by looking in the same direction away from the centre of the expansion (the big bang) or looking back at the centre, or for that matter any other directions? I believe that the results may show that the galaxies are moving away from the centre of expansion and each other at different speeds.[5]

If the big bang really had been an exploding ball of matter, then some galaxies would lie deep in the midst of the mêlée, surrounded on all sides, while others would be located near the edge of the assemblage. Suppose this were so, and picture the view from a far-flung galaxy. In one direction would lie the centre of the universe; in the opposite direction there would be empty space. The sky would appear dramatically different depending on which way an observer looked. That is certainly not what we see from Earth: the universe looks very much the same in all directions. As far as our telescopes can penetrate, which is about 13 billion light years, encompassing roughly 100 billion galaxies, matter is distributed uniformly (strictly, it is *clusters* of galaxies that are distributed uniformly). There is no evidence for any bunching up around some sort of centre or, conversely, for any thinning out towards an edge.

How, then, should we describe the big bang and the expanding universe, given these observational facts? Suppose that the universe were frozen as it is today, with galaxies distributed evenly throughout space (on average). Now imagine that we start playing the cosmic movie, watching the universe expand. Taking a god's-eye view, there would be no systematic flow of galaxies away from *any* particular point in space. Rather, all the clusters of galaxies move away from all the others at the same rate. Everywhere you look, the gaps between the clusters of galaxies get progressively bigger: there is more and more space surrounding each cluster as time goes on. An observer in any given galaxy will *seem* to lie at the centre of a pattern of expansion, because all the other clusters of galaxies are moving away, but they all are moving away from one another too, so the observer's impression of being located at the centre is illusory. There is no centre. For this reason, the terms 'explosion' and 'big bang' are rather inappropriate, although we are probably stuck with them now.

Confirmation of this no-centre-no-edge picture comes from the CMB. If the big bang had happened at a given point in space, then that part of the sky would be aglow with primordial radiation, whereas the sky facing away from the centre of the big bang, towards the void, would be cold. In fact, as I have explained, the background radiation is uniform across the sky, when the tiny temperature variations are averaged out: there is no hint of it being systematically hotter in one part of the sky than in the opposite part.

Expanding space

Cosmologists have struggled to find ways to describe the expanding universe in simple language. Here are some popular attempts:

Space is *in* the universe rather than the universe being in space.

The big bang happened everywhere, not at one point in space.

The big bang was the explosion *of* space, not an explosion *in* space.

A simple analogy that may help is to imagine a very long string of elastic with beads attached at regular intervals (see Figure 5). As the elastic is stretched, so the beads move apart. Every bead extends its separation from its neighbours, so the view from any given bead will

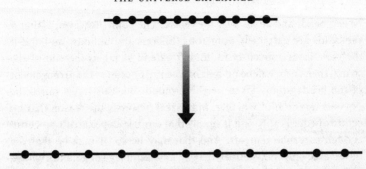

5. The expanding universe

In this one-dimensional analogy, the expanding universe is represented by an elastic string (standing for space) to which beads (standing for galaxies) are attached at intervals. As the elastic stretches, so each bead moves away from all the others. The view from any given bead is the same: the other beads all seem to be moving away at speeds proportional to their distances. So long as we can see neither end of the string, all beads seem equivalent: there is no centre and no edge apparent from the distribution or pattern of motion.

be of other beads moving away. All beads are equal: there is no central bead.

The analogy works better if you envisage space itself to be elastic, and capable of being stretched. In fact, the idea of space stretching is not just an analogy: it comes close to what physicists believe to be the case. The ability of space to stretch, as well as to become curved or warped, is the basis of Albert Einstein's general theory of relativity. Looked at in this way, the expansion of the universe is not so much the migration of galaxies *through* space as the stretching or swelling of the space *between* the galaxies, an image that leads to yet another popular description:

The expanding universe is like a currant bun swelling in the oven, with the currants playing the role of galaxies and the dough representing space.

The attentive reader will notice a fudge in the elastic string analogy, because real strings of elastic are finite in length, so there will always be end beads and beads near the centre. The point, however, is that if the beads are extremely numerous (billions and billions, say) and if the view from a given bead doesn't extend as far as the end of the string, then there will be no hint of a centre or edge in the arrangement of the beads in any given neighbourhood. But this fudge raises the obvious question of whether, in the real universe, the reason that no centre or edge is apparent is simply that our telescopes aren't powerful enough to probe that far. And that may be so. It may be that the observable universe is buried deep in an assemblage of galaxies which, viewed on a grander scale, *does* have an edge. (It may not have an obvious centre unless the assemblage is roughly spherical in shape.) If the edge is far enough away, then the lopsidedness of the heat radiation would not yet be apparent because there would not have been enough time since the big bang for the final, dwindling heat radiation, travelling at the speed of light, to have reached us from the edge.

On the other hand, the universe may not be like that at all. It may be infinite in all directions. I well remember when I was about eight years old asking my father where space ended. He replied that space cannot have an end, for if it did, then it would raise the question of what lay beyond the farthest point. It is an old argument, but as I shall explain later, it may in fact be wrong. To be sure, there is no logical reason why space cannot be infinite and populated everywhere by galaxies. There would then be an infinite number of galaxies, spread uniformly in clumps as we see them, extending through infinite space for ever and ever. This is the simplest, default assumption: what you see is what you get, *everywhere*. In discussions of the structure of the universe, this uniformity assumption is referred to as the *cosmological principle*. It is an application of a more general principle, called the *principle of mediocrity* – that there is nothing special or privileged about our location in the universe.

At the risk of confusing you, I should say that the cosmological principle does not necessarily imply an infinite universe. It is possible for the universe to be finite in volume without possessing a centre or an edge. These apparently contradictory properties can arise if we allow for the possibility of space being warped. Let me explain this by extending the elastic string analogy. Suppose the string were in fact a giant circular elastic band, expanding to a larger and larger radius

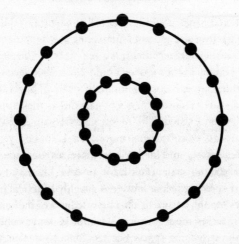

6. Closed space

It is possible for three-dimensional space to be finite. The circular elastic band captures in one dimension the quality of space being closed, yet having no centre or edge (i.e. there is no centre or edge at any point on the band). As the 'universe' expands, the band stretches to a larger radius without upsetting the no-centre-no-edge property.

(see Figure 6). The beads are attached all around the band at regular intervals. There are neither end beads nor a central bead: all beads are truly equal, yet all beads move away from all others as the elastic loop is expanded. This is an example of a finite yet unbounded bead universe, made possible by allowing the elastic 'space' to be closed into a loop. Shortly I shall explain how real three-dimensional space might be closed in this manner. But first there is a more pressing issue to deal with.

The speed of light and the view from planet Earth

When astronomers peer at the heavens through telescopes they see distant objects not as they are now, but as they were when the light reaching the telescopes embarked on its journey across space. In this respect, a telescope is also a 'timescope'. For example, if a nearby star exploded yesterday we would be blissfully unaware of this cataclysm for years, until the pulse of light announcing the star's demise arrived on Earth. Looking farther afield, we see stars in the neighbouring Andromeda Galaxy as they looked about $2^1/2$ million years ago. More distant galaxies appear correspondingly older. The Hubble Space Telescope routinely records images of galaxies as they appeared long before Earth even existed. The oldest galaxies can actually be seen still in the process of formation, more than 12 billion years ago. So by penetrating farther and farther into space, astronomers can watch the history of the universe unfolding in reverse. Light may travel fast, but its speed is nevertheless finite – a fact that has very important consequences for the nature of the universe, as we shall now see.

As light traverses the expanding universe its wavelength stretches along with the stretching space. Because longer wavelengths of light appear redder in colour, the effect is referred to as the *red shift*. It was the fact that more distant galaxies look redder than nearby galaxies that first alerted Edwin Hubble to the expansion of the universe in the 1920s. It is the same effect that brings about the cooling of the cosmic background heat radiation.

The amount of red shift depends on how long ago (and hence how far away) the light was emitted. Working back towards the big bang, the red shift gets bigger and bigger. It is not at all uncommon these days to see photographs of galaxies (or quasars – the energetic hearts of some violently disturbed galaxies) whose light has been stretched to two or three times their original wavelengths, so the distance between us and those galaxies will have substantially increased in the time it has taken the light to reach us. Going much farther back in time (and out into space), we reach the epoch from which the CMB emanates. The temperature at that time was about 3,000 K, hot enough to radiate in the ultraviolet region of the electromagnetic spectrum (see Figure 7). It corresponds to a red shift of about 1,100, which is big enough to stretch the wavelength all the way from ultra-violet, through visible light and infrared, into the microwave region of the spectrum. So what was emitted as (hot) ultraviolet radiation appears to us today as (cooler) microwaves because of the enormous factor by which the universe has expanded in the interim.

As I have mentioned, the CMB has travelled to Earth relatively

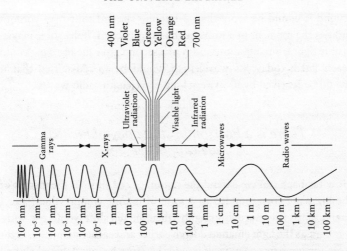

7. The electromagnetic spectrum

Electromagnetic waves can have any wavelength. Although they are the same phenomenon, the names we give to them depend on the wavelength – shown here in units ranging from nanometres (nm) through micrometres (μm) to kilometres (km). The red shift due to the expansion of the universe causes radiation emitted in the ultraviolet region of the spectrum at 380,000 years after the big bang to have its wavelength stretched into the microwave region by the time we receive it.

undisturbed since about 380,000 years after the big bang. Before that time the temperature was too high for atoms to exist because the electrons would have been stripped away from the nuclei by the intense heat, i.e. the atoms were ionized. Physicists refer to a gas in this state as plasma. Plasmas scatter light strongly and so they are opaque: that is why we can't peer inside the sun, for example, through the surface glare. Therefore, when WMAP detects the CMB, it is in effect seeing as far back in time as is possible with electromagnetic waves.[6] No ordinary telescope or microwave antenna, however powerful, can penetrate the glowing fog beyond. Nevertheless, we can pretend that the gas isn't there, and calculate what the red shift would have been at earlier epochs if we had an unrestricted view. Ten seconds after the big bang, the red shift would be roughly a billion; at one second it would be more than 3 billion. The red shift rises without limit as the moment of cosmic birth is approached. If, by some magic, visible light from one-thousandth of a second after the big bang could reach Earth today, its wavelength would be so red-shifted that it would be received by us as very long wavelength radio waves.

There is a horizon in space beyond which we cannot see

How far back can we extend the notion of red shift? If the instant of the big bang corresponded to a state of infinite compression (later I shall discuss this more thoroughly) then the red shift would rise without limit as the light emanated from earlier and earlier moments. Light from the big bang event itself would therefore be infinitely red-shifted – if it could ever reach us, that is. An infinite red shift implies that we would actually see nothing at all: the radiation would convey no information. This, clearly, is a fundamental limit: we could not see beyond this point in space or this moment in time. Cosmologists refer to the limit as a *horizon*. The moment of the big bang, in this very simplified and idealized picture, is a horizon in space beyond which we can never see, even in principle, however powerful our instruments (and ignoring the opacity of the material). One way of expressing this restriction is to point out that in the 13.7 billion years that have elapsed since the big bang, light can have travelled at most 13.7 billion light years, and it is therefore no surprise that we cannot see beyond this distance. Such statements must, however, be treated with care. When we see a galaxy 13 billion light years away, we see it where it was located 13 billion years ago. Today, this galaxy will be much farther away from us because the universe has expanded greatly in the intervening time. So the distance to the galaxy depends on whether we are talking about the look-back distance (i.e. the distance to where the galaxy *was*, when its light was emitted, looking back in time) or the current distance. The farthest galaxies that we can see using the Hubble Space Telescope are *now* located about 28 billion light years from Earth.[7]

Another confusion often surfaces at this stage of the discussion.

People will ask why, if the universe started out extremely shrunken in size, we have to wait so many years for light to reach us from regions of the universe that were, in the far past, located much closer. Why didn't light from those regions quickly traverse the compressed universe and arrive in our region long ago? The answer to that question lies with the *rate* of cosmic expansion. As light moves across the universe, it is chasing after galaxies that are receding from the light's source. Expressed differently, even as light crosses space, space itself expands ahead of it, so a pulse of light resembles a runner on a treadmill. As a result, the journey time becomes greatly extended.

The horizon has a significance beyond merely truncating our view of the universe. It is a fundamental tenet of the theory of relativity that no object or physical influence can exceed the speed of light (see Box 1). Therefore an event that occurs beyond our horizon not only cannot be seen, it cannot exert any physical effect on us at this time, and vice versa. As we shall see, this restriction on the operation of cause and effect turns out to be an important factor in our attempts to understand the structure of the universe.

Different definitions of 'the universe'

The existence of a light horizon introduces a great deal of muddle into what is meant by the word 'universe'. Some authors are lamentably vague about the term; I need to be more precise. So here goes.

The observed universe

This is all of space and its contents out as far as our instruments can currently probe. A century ago, the observed universe consisted of little more than our galaxy and its near neighbours, but with the development of bigger telescopes astronomers can now see almost as far as the horizon. These days the observed universe virtually coincides with the next definition:

1. Why light sets the cosmic speed limit

Why can nothing travel faster than light? One way to approach this question is to ask what happens if you try to make a physical object break the light barrier. This experiment can be done with charged subatomic particles such as electrons, which can be sped up in particle accelerators to speeds approaching that of light.

According to Einstein's special theory of relativity, published in 1905, the mass of a particle depends on its speed. This is a consequence of the famous $E = mc^2$ formula, where c denotes the speed of light. Associated with motion is a form of energy (called kinetic energy), and the formula tells us that energy (E) has mass (m). Because a moving object has more energy than a static one, it must have more mass. At everyday speeds we don't notice that moving objects are heavier than static ones – the factor c^2 in the formula implies that everyday amounts of kinetic energy contribute only a minuscule additional mass. However, at speeds close to that of light the mass of the kinetic energy rivals that of the mass at rest. At 87 per cent of the speed of light, the kinetic energy weighs more than the rest mass. At still faster speeds, the total mass starts to escalate. What then happens is a case of diminishing returns. More and more of the effort expended in trying to accelerate the particle goes into making it heavier, and less and less into making it move faster. As light speed is approached, the mass of the particle rises without limit, making it impossible to increase its speed any more. The light barrier is therefore unbreakable.

Now for a bit of terminology. The mass of a particle at rest is called, naturally enough, its *rest mass*. The total mass of a moving particle is called its *relativistic mass*, and consists of the rest mass plus the mass associated with the energy of motion. When physicists refer simply to 'the mass' of a particle, they usually mean the rest mass. A photon is said to have zero rest mass, but it can never be at rest: it travels always at the speed of light. It certainly has non-zero relativistic mass (proportional to its frequency, in fact).

The blithe statement that 'nothing can travel faster than light' is actually a bit misleading. Special relativity forbids a material object from passing another at a speed faster than the speed of light. But special relativity is a restricted part of the general theory of relativity: the latter takes gravitation into account and permits such things as the expansion of space. In these circumstances, the no-faster-than-light rule is transcended. Distant galaxies, for example, can be effectively retreating from us at faster than light speed. This does not contradict the rule as applied within special relativity, which refers to a local situation and not to the global motion of the universe.

The observ*able* universe

Taking into account that we cannot see beyond the horizon, the observable universe means 'everything within the horizon'. Over time, the horizon expands: in another billion years it will be 14.7 billion light years in radius. So what is meant by the observable universe depends on when you are looking at it.[8] The fact that there is a horizon doesn't mean that there is emptiness beyond.[9] You can envisage the horizon as the surface of an imaginary sphere 13.7 billion light years in radius, centred on Earth, moving out in all directions at the speed of light and determining the limits of how far we can see, even in principle. There is nothing special in this respect about Earth: each point in the universe will have its own spherical horizon around it, perhaps overlapping ours. Notice, however, that if you were to be instantly transported to a distant galaxy X, say 8 billion light years from Earth, then the horizon around X would extend into regions of space that we here on Earth cannot see at this time. We don't know what lies over our cosmic horizon, but it seems reasonable to suppose that it's more of the same, and that the view from galaxy X should not therefore be a lot different from ours. The simplest assumption is that the region of space within our horizon is typical of the entirety: this is the principle of mediocrity again. If so, then we can formulate a third definition:

The entire universe

This includes all of (possibly infinite) space, within and beyond our horizon, plus all of its contents, on the said assumption that the observed universe is typical of the whole. Later we shall see that this simple view of the universe, based on an uncritical application of the principle of mediocrity, is now being challenged, leading to yet another definition:

The pocket universe[10]

This is the region of space out as far as it resembles the observable universe we see today (which may extend a very long way indeed – far beyond our horizon). But if we inhabit a pocket universe there would be a boundary somewhere, far, far away, beyond which things would look very different. There would, however, be other pocket universes scattered far and wide in this region beyond, some resembling ours, but most of them not. Therefore, our pocket universe would probably be very atypical, implying that the principle of mediocrity fails if we consider the entire collection of pocket universes. This leads to the final definition:

The multiverse

Roughly speaking, this is the collection of all pocket universes (probably infinite in number) plus the gaps in between. Some authors prefer the term 'megaverse'. I shall have much to say about the multiverse in later chapters.

Warped space

So far I've been concentrating on astronomical discoveries. But cosmology would not be a true science if it lacked a framework of physical theory within which these discoveries can be understood. The theoretical basis of modern cosmology was established almost a century ago by Einstein in the form of his general theory of relativity.

The theory was published in 1915, in the dark days of World War I, but this did not stop astronomers and physicists from both sides of the conflict taking a keen interest in what it had to say about cosmology. The general theory of relativity, or general relativity as it is normally abbreviated, was designed to replace Newton's seventeenth-century theory of gravitation. In cosmology, gravitation is the dominating force, overwhelming all others because of the vast mass of the cosmos, so it is to gravitational theory that cosmologists turn to understand the expanding universe.

Einstein's brilliance was to spot that although gravitation manifests itself as a force, it may also be understood in a completely different way, in terms of 'warped geometry'. Let me explain what this means. The rules of geometry we learn at school date from the time of ancient Greece: the subject is often referred to as Euclidean geometry, after Euclid, who wrote it all down. There are many well-known theorems that can be proved from Euclid's axioms, for example the famous one named after Pythagoras. Another well-known theorem is that the angles of any triangle add up to two right angles (180°). These properties of lines, circles, triangles, and so on are watertight, but they come with one important proviso: they apply to *flat* surfaces. The theorems work correctly on blackboards and sheets of paper on school desktops, but they do not work on curved or warped surfaces such as globes. Pilots and navigators are well aware of this, and they have to use different geometrical rules to cope with the Earth's curvature. For example, on the Earth's surface a triangle can contain *three* right angles (see Figure 8).

If two-dimensional surfaces can be either flat (Euclidean geometry) or warped (non-Euclidean geometry), could *three*-dimensional space also have either 'flat' (Euclidean) geometry or warped geometry? Before Einstein, almost everyone assumed that space had 'flat' or Euclidean geometry, straightforwardly extended from the rules we learn for two dimensions. But there is no logical reason why that must be so. Some nineteenth-century mathematicians toyed with the idea that the geometry of three-dimensional space could be a generalization of curved surface geometry. They worked out the geometrical rules for this 'warped space', but at the time it was treated simply as a mathematical game. All that changed with general relativity. Einstein

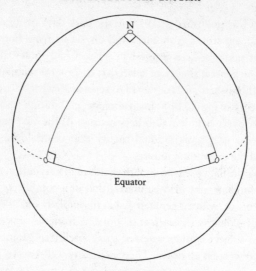

8. Warped space

On a spherical surface the rules of geometry differ from those on a flat sheet. For example, a triangle may contain three right angles – such as this triangle on Earth's surface, with its apex at the North Pole and its base along the equator. The two-dimensional spherical surface represented here is an analogue of warped three-dimensional space.

proposed that a gravitational field can warp three-dimensional space, necessitating the use of non-Euclidean geometry to describe it.

What, then, is curved space? One way to imagine it is to think of a triangle drawn around the sun (see Figure 9). Importantly, this must be a *flat* triangle (i.e. it lies in a plane). Now measure the angles and add them up. If Euclid's geometry applies to this situation, the result will be 180°. Einstein, however, claimed that the answer should be slightly greater than 180°, even though the triangle is flat, because the sun's gravitational field warps the three-dimensional geometry of the space around it. This experiment can actually be done (more or less) by bouncing radar waves off Venus and Mercury and doing triangulation. It turns out that Einstein is right – space really is curved rather than flat. (There is an important terminological nicety here: when cosmologists talk about 'flat' space, they don't mean space flattened

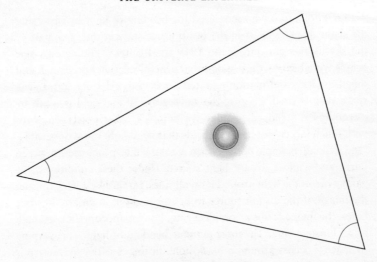

9. The sun's spacewarp

If a *flat* triangle is drawn round the sun, the angles add up to a bit more than 180° because the sun's gravitational field distorts the geometry of space in its vicinity. An equivalent way to think about this phenomenon is that the sides of the triangle are the straightest lines possible in the curved geometry. If light beams were directed along the sides of the triangle, it would seem to the receiver on the far side of the sun that the beams had been slightly bent by the sun's gravity.

to a pancake, they mean three-dimensional space with Euclidean geometry.) Sometimes the curved geometry near the sun is described by saying that the sun's gravity bends light rays passing near it, in which case the triangle would have distorted angles because the sides are wonky. This is true: it is an equivalent way to think about curved space, with the important point that the wonky sides are actually the straightest possible lines that can be drawn in the warped geometry, so it isn't just a matter of straightening the bent light beams out and recovering Euclid's results. The space is irreducibly curved, and no amount of manipulation will make it conform to Euclidean rules.

The warping of space around the sun, although detectable, is nevertheless tiny. Its existence was confirmed by the English astronomer

Arthur Eddington, who measured the bending of light by observing the slight displacements in the positions of stars in the same part of the sky as the sun during the 1919 total eclipse. Eddington's star beams were bent by the amount that general relativity predicted, and this dramatic confirmation elevated Einstein to celebrity status. The spacewarp is small because the sun's gravitational field is weak by astronomical standards. Today we know of other objects in space with much larger gravitational fields that bend light more noticeably. One striking example occurs when a galaxy interposes itself between Earth and a more distant light source. Under these conditions, the galaxy bends the light around it on all sides, rather like a lens, causing the image of the distant source to be smeared out in an arc. In some cases, the image forms a complete ring, known appropriately enough as an Einstein ring. The most extreme bending of light – or warping of space – occurs around a black hole. In this case the spacewarp is so strong that it actually traps light completely, preventing it from escaping.

I have simplified the foregoing account in one important respect. In his earlier, so-called special theory of relativity, published in 1905, Einstein demonstrated that space is linked to time in a manner that makes it natural to consider the whole package – *spacetime* – together. Space has three dimensions and time has one, making four dimensions in all.[11] Hermann Minkowski, one of Einstein's mathematics teachers, worked out how to modify the rules of Euclidean geometry to describe four-dimensional spacetime. When Einstein went on to generalize his theory of relativity in 1915 to include gravitation, he proposed that it is spacetime that is warped, and not merely space. Distorted spacetime geometry may imply warped space, warped time, or both. In the foregoing discussion of the spacewarp around the sun, I have ignored the time aspect. That is important too, and the sun's (tiny) timewarp has also been measured. In fact, Earth's even smaller timewarp is measurable; it manifests itself by the fact that clocks tick very slightly faster at higher altitudes – on a mountain top, say – than at sea level. (For more on tests of general relativity, see Box 2.)

2. Tests of Einstein's general theory of relativity

When Einstein published his general theory of relativity in 1915 (building upon the special theory of 1905) he proposed three tests which could be performed by making observations from within the solar system. The first of these was the bending of starlight, described on p. 42. The second test had to do with the orbit of the planet Mercury. The warped spacetime around the sun causes the orbits of the planets to execute a slow twisting motion, technically known as precession of the perihelion. A similar, but much larger precession is caused by perturbations by other planets. For Mercury, the predicted small correction due to relativity is 43 arc seconds per century. Although tiny, it had already been measured by astronomers and its cause was something of a mystery. General relativity explained it exactly.

The third test concerned the effect of gravitation on time: clocks run slower in a gravitational field. In fact, all physical processes are slowed, and this includes the emission of light. Einstein pointed out that light from the sun should be shifted slightly to the lower-frequency end of the spectrum as a result of the sun's gravitational field, an effect known as the gravitational red shift. This is hard to measure for the sun, but in 1960 the prediction was verified quite accurately by utilizing Earth's gravity. The experiment consisted of sending gamma-ray photons up a tower at Harvard University and using an ultra-sensitive nuclear resonance technique to detect the tiny shift in frequency.

Advances in astronomy have revealed objects in space with huge gravitational fields, for which the effects of general relativity are far greater than they are within the solar system. For example, pairs of neutron stars in close orbit have been used to confirm many of the theory's predictions to high accuracy. Black holes are another extreme prediction of general relativity, and their properties are being actively studied by astronomers. Gravitational lensing (see p. 42) provides an additional test.

With the invention of atomic clocks and lasers, it has become possible to test the predictions of general relativity to greater and greater accuracy and in many more situations. Indeed, relativistic effects have now entered the world of practical engineering. The Global Positioning System used by drivers and pilots operates by comparing accurately timed signals from a network of orbiting satellites, and the computer algorithm must factor in the time-warp effects of both gravitation and motion.

One prediction of the general theory of relativity remains incompletely tested. In 1918 Einstein showed how moving masses can generate gravitational waves. These are analogous to electro-magnetic waves, and also travel at the speed of light. However, because gravitation is so much weaker than electromagnetism, the effects of gravitational waves are extremely small. One way to envisage them is as ripples in the curvature of space, so that if a gravitational wave passes by it will cause all bodies to vibrate very slightly. Detectors have been built to try to spot this effect, but so far without success. However, a binary neutron star system that has been studied for many years shows clear signs of orbital decay due to the emission of gravitational waves, so their exist-ence has been confirmed indirectly.

Einstein's finite but unbounded universe

The observable universe contains about 10^{50} tonnes of visible matter in the form of stars, gas and dust, all of which combines to create a powerful gravitational field. Because gravitation warps the geometry of space, an interesting question immediately arises: what is the overall shape of the universe? By this, I don't mean how the galaxies are distributed in space. What I am referring to is *the shape of space itself*, considered on the grand scale of the cosmos. This was the problem Einstein set out to address in 1917, two years after he first presented his general theory of relativity. By applying the idea of warped space to cosmology he was able to construct a mathematical model of the entire universe. Although the model turned out to be rather wide of the mark, it introduced into cosmology several important features.

As I have explained, the sun creates a small distortion of space in its vicinity. Others stars create similar localized distortions. The question I am now considering is how all these distortions combine together. Will the curvature be cumulative, so that when we come to consider clusters of galaxies the spacewarp will be getting seriously big, or will the distortions tend to cancel each other out? In Einstein's mathematical model of the universe the curvature accumulates so that, averaged out over billions of light years, the shape of space resembles a three-dimensional version of the surface of a sphere, which is referred to as a hypersphere. Don't worry if you can't envisage a hypersphere. The important point is that it makes good mathematical sense, and its properties are easy to calculate by generalizing the geometry of familiar two-dimensional spherical surfaces.

An important property concerns the volume of space. In Einstein's hyperspherical universe space is finite (just as the Earth's spherical surface is finite). This means that space (in Einstein's model) does not extend for ever – thus contradicting what my father taught me. Another important property of Einstein's universe is that it is uniform (on average). The same is true, of course, of the surface of a sphere. There are no distinguishing features that single out any particular spot on a spherical surface as special; there is no centre or boundary. (The Earth has a centre, of course, but the *surface* of the Earth has no centre.) So Einstein's universe would look the same from any galaxy, precisely as astronomers observe. It is therefore finite yet unbounded – unbounded in the sense that there is no edge or barrier to prevent an object travelling from one place to any other place in the universe. Yet there are a limited number of places to go, in the same sense that there are a limited number of places to visit on the Earth's surface. And just as one can circumnavigate the Earth by always aiming straight ahead – returning home from the opposite direction – so one could in principle go round the Einstein universe, by aiming in a straight line, never deviating and returning from the opposite direction to that in which you had set out. Indeed, with a powerful enough telescope, you could look right around the Einstein universe and see the back of your own head! This model of space is the three-dimensional generalization of the elastic loop that I described earlier in the chapter (Figure 6, p. 31).

One of the difficulties people have in conceptualizing a hypersphere is with the troublesome issue of 'what lies in the middle'. They think of a spherical two-dimensional surface, such as a round balloon, and say, 'well, the balloon has air inside it'. The issue of what the Einstein universe 'encloses' is a bit of a red herring, however. We humans, and the universe we perceive (at least those parts of it we have so far perceived), are restricted to three dimensions of space, so the issue of what, if anything, lies 'inside' Einstein's three-dimensional hyperspherical space is moot. If it helps, you can envisage this 'interior' region as a fourth dimension of space (empty, or filled with green cheese for that matter), but because we are trapped in the hyperspherical three-dimensional 'surface' it doesn't make a jot of difference to us whether the interior region is there or not, or what it contains. Much the same goes for the exterior region, the analogue of the space outside the balloon.[12]

To ram this point home, since it proves so hard for people to grasp, try to put yourself in the position of a pancake-like creature restricted to life on the surface of a round balloon. The pancake might conjecture about what lies inside the balloon (air, empty space, green cheese . . .), but whatever is there doesn't affect the pancake's actual experiences because it cannot access the space inside the balloon, or receive any information from it. Furthermore, it is not even necessary for there to be *anything at all* (even empty space) inside a spherical surface for an inhabitant of the surface to deduce its sphericity. That is to say, the pancake doesn't need a god's eye-view of the balloon to conclude that its world is spherical – closed and finite, yet without boundary. The pancake can deduce this entirely by observations it can make from the confines of the spherical surface: the sphericity is *intrinsic* to the surface, and does not depend on it being embedded in an enveloping three-dimensional space. How can the pancake tell? Well, for example, by drawing triangles and measuring whether the angles add up to more than 180°. Or the pancake could circumnavigate its world. In the same vein, humans could deduce that we are living in a closed, finite, hyperspherical Einstein space without reference to any higher-dimensional embedding or enveloping space, merely by doing geometry *within* the space. So the existence or otherwise of an 'interior' or 'exterior' region of the Einstein universe, not to mention what it

46

consists of, is quite simply irrelevant. But if you would like to imagine inaccessible empty space there for ease of visualization, then go ahead. It makes no difference.

What shape is the universe?

All this is well and good, but was Einstein actually right about the universe being shaped like a hypersphere? Here, WMAP has been of immense help. Obviously, if the universe were seriously lopsided, it would show up in the pattern of microwaves from the sky. The fact that this radiation is so uniform already indicates that the universe, out as far as we can see, is at least fairly regular in shape.[13] But what shape is it? Resorting again to a two-dimensional analogy, we can immediately identify two perfectly regular shapes: an infinite flat sheet and a perfect sphere. But there is a third shape, a sort of inverse of the sphere. Remember that on a sphere a triangle has angles adding up to more than 180°. Technically, the sphere is defined to be curved *positively*. What about a uniform surface on which the angles add up to *less* than 180°? This is a space with *negative* curvature. Such a surface exists, and it looks a bit like a saddle, but infinitely extended (see Figure 10). All three surfaces – with zero, positive and negative curvature – can be generalized to three dimensions. Since the 1920s, when cosmologists first realized that there were three different shapes for uniform space, they have wanted to know which one our universe most closely resembles.

A direct assault on the problem has been tried many times. Because the geometry of the three different spaces is different, astronomers ought to be able to tell simply by looking. Measuring the angles of a triangle over cosmic distances isn't feasible, but there are other possibilities. Returning again to two dimensions, imagine drawing a series of concentric circles on a flat sheet. The area enclosed by each circle rises in proportion to the square of the radius: double the radius and the area is four times as great. But on the surface of a sphere this relationship goes wrong: the area increases with radius *less* rapidly. That's easy to see, because if you tried to flatten a cap, you would have to cut wedges out of it, so it would fail to cover a disk of

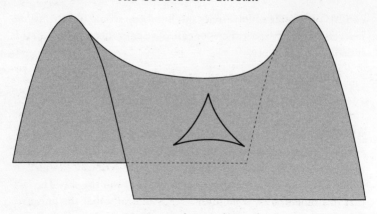

10. *Negatively curved space*

It is possible that, on the cosmological scale, space is uniform but curved outwards instead of inwards. The two-dimensional surface shown here is the analogue of such a negatively curved, three-dimensional space. It is infinite and homogeneous. The negatively curved geometry manifests itself in the distortion of a triangle, whose angles add up to less than 180°.

equivalent radius on a flat sheet. Similarly, the area on the saddle shape increases *more* rapidly than the square of the radius. Converting all this into three dimensions means that the volume of a region of space will rise as the cube of the radius if the universe is flat (three-dimensionally flat, remember, not pancake-flat). If the universe is a hypersphere, as Einstein suggested, the volume will increase less rapidly with radius, and if it is a 'hypersaddle' it will increase more rapidly. The volume of a region of space can be assessed by counting the number of galaxies it contains.

Some astronomers tried to establish the geometry of the universe in this direct way, but their results were inconclusive on account of the difficulty of measuring precise distances to far-flung galaxies, and other technical complications. However, the answer can be inferred from the WMAP data, by measuring the sizes of the temperature fluctuations – the hot and cold (light and dark) splodges in Figure 3 (p. 21). Before WMAP was launched, theorists had already worked out how big the physical sizes of the strongest fluctuations should be. Converting that into apparent angular size in the sky depends on the geometry of space: if the universe is positively curved it would make the angles appear larger, while negative curvature would make them smaller. If the universe is geometrically flat (i.e. has Euclidean geometry), the angular size of the strongest hot and cold fluctuations should be about 1° across. The results that flowed back from the satellite were definitive.[14] The fluctuations were very close to 1° in size, a result confirmed by ground-based and balloon-based experiments. Cosmologists then declared that, to within observational accuracy of about 2 per cent, space is flat.[15]

The universe might weigh nothing!

How can space be flat overall when the sun and stars distort it locally? Obviously something in between the stars must make space curve the other way to average out to zero. (Remember that space can curve both positively and negatively.) What is this something? The answer comes from Einstein's famous equation $E = mc^2$, which tells us that mass is energy, and energy has mass. I shall often refer to 'mass-energy' as a single concept in what follows. In estimating the mass of the universe we have to include all forms of energy, not just the mass of matter in the sun, the stars and other astronomical objects. Also contributing to the total mass-energy are the heat energy of the CMB, magnetic fields and cosmic rays. Last, but by no means least, is the gravitational field itself: gravity is a form of energy. But now we notice a curious fact. Imagine trying to pluck the Earth out of its orbit around the sun. You would have to do work – that is, expend energy – to draw it away against the sun's gravitational pull. So the gravitational energy binding the Earth to the sun is *negative* (it requires work to sever the bond). If the gravitational field has negative energy, it must also have negative mass, and must be *subtracted* from the positive mass-energy of the sun and planets.

Let's see how the negative gravitational energy comes into the total mass-energy of the universe. Within the solar system the amount of gravitational mass-energy is puny compared with the enormous mass of the sun. The overall mass of the solar system, even with all the gravitational binding energy included, is still large and positive. But when it comes to the universe as a whole it's a different matter. One of the distinctive things about gravity is that it is universal: it acts between all particles of matter in the universe. So to work out the (negative) gravitational energy for the whole universe you have to tot up all the gravitational energy due to *every* object tugging on *every other* object: that's a lot of tugs in all, even if for each star the vast majority of the others are enormously distant. A simple estimate of the gravitational energy binding all the galaxies to each other gives an effective mass for the gravitational field (using $E = mc^2$) of about *minus* 10^{50} tonnes, which is roughly equal (and opposite) to the mass of all the stars and other stuff. The fact that the two very large numbers are of the same order, and of opposite sign, suggests very much that they are doing their best to cancel each other and make the net mass of the universe zero!

Einstein's general theory of relativity provides a link between the mass of the universe and the geometry of space (see p. 140). Specifically, if the total mass is positive – matter wins out over negative gravitational energy – then space is curved positively, like Einstein's universe. If the mass is negative – gravitational energy wins out over matter – then space is curved negatively, like the saddle. If it is precisely zero then space is flat.[16] Cosmologists knew for years that the positive and negative contributions to the mass of the universe roughly cancel out. But WMAP clinched it. To within the 2 per cent accuracy of the measurement, the satellite found space to be flat, which translates into the conclusion that *the universe contains no net mass at all!* And that, as we shall see later, is yet another one of those 'coincidences' that is needed for a life-permitting universe.

How many dimensions are there?

In 1884 the splendidly named English clergyman Edwin Abbott Abbott published the revised edition of his book *Flatland*, which was destined to become a classic, and is still widely read today.[17] The author prefaces his work with a curious dedication:

To
The Inhabitants of SPACE IN GENERAL
And H. C. IN PARTICULAR
This Work is Dedicated
By a Humble Native of Flatland
In the Hope that
Even as he was Initiated into the Mysteries
Of THREE Dimensions
Having been previously conversant
With ONLY TWO
So the Citizens of that Celestial Region
May aspire yet higher and higher
To the Secrets of FOUR FIVE OR EVEN SIX Dimensions
Thereby contributing
To the Enlargment of THE IMAGINATION
And the possible Development
Of that most and excellent Gift of MODESTY
Among the Superior Races
Of SOLID HUMANITY

For those readers who aren't familiar with *Flatland*, it is a story about life in two dimensions, and the bewilderment of two-dimensional creatures, unable to envisage 'up' and 'down', having to confront processes in three dimensions. The implication of this story is that we human beings are equally bewildered when trying to imagine a world with more than three dimensions (at least, this human being is). But the mere fact that we find something hard to visualize is no argument against its being correct. I find it hard to envisage a billion dollars, but apparently some people actually possess such sums of money.

For Abbott, the possible existence of extra dimensions of space was merely an amusing diversion. He had no evidence that space actually possesses more than three dimensions and no reason, based on the physics of his day, to postulate such a thing. (I am not referring to the fact that time is sometimes described as the fourth dimension, which is different entirely.) But forty years after Abbott's book appeared, the possibility that space might actually have *four* dimensions rather than three was seriously suggested by a German mathematician named

Theodor Kaluza. Before I get into Kaluza's theory, I should explain what I mean by an extra dimension of space. Familiar three-dimensional space permits movement in three mutually perpendicular directions: for example, up-and-down, back-and-forth and side-to-side. Any movement must involve a displacement along one, or a combination, of these three directions. There is simply nowhere else to go. A fourth dimension would allow a *fourth* direction, lying perpendicular to the other three. Clearly you cannot have that in familiar space, but one can study a fictional space with such a property. It isn't much harder to do the same for five, six or more dimensions; scientists and engineers routinely use such constructions in their work, as a helpful computational device.

So now we get to the main issue. Might the universe actually have four (or even more) space dimensions? The knee-jerk answer is no, because we can see it hasn't. But let's not be too hasty. The fourth dimension might be there all right, but hidden from us in some way.

Hiding dimensions of space

How do you hide a dimension of space? There are two different ways to do it. The first was suggested in the 1920s by the Swedish physicist Oskar Klein. His idea is very simple. Imagine viewing a garden hose from a distance; it looks like a wiggly line. On closer inspection, however, the line turns out to be a two-dimensional sheet rolled into a thin tube (see Figure 11). What at first appears to be a point on a line is in reality a little circle around the circumference of the tube. In the same way, perhaps what we take to be points of three-dimensional space are really little circles going around a fourth dimension. If the circumference of these circles were much smaller than an atom we wouldn't notice the presence of the extra space dimension from casual observation; it would be revealed only from careful experimentation in subatomic physics.

There is no limit to the number of extra dimensions that can be rolled up – or 'compactified', to use the official jargon – in this manner. But with more than one extra dimension there is an increasing variety of ways to compactify. For example, two extra dimensions could each

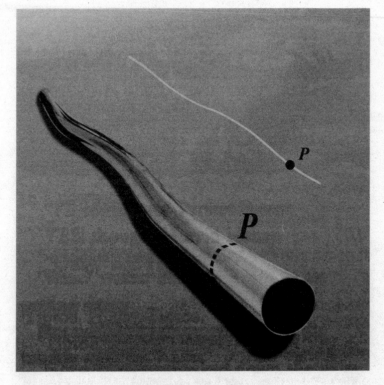

11. *How to hide a space dimension*

From a distance, a garden hose looks like a wiggly one-dimensional line, but on closer inspection we see that it is really a thin tube. The point P is actually a little circle going around the extra, second dimension. Space could be like this: each point in space might, at sufficiently high magnification, be revealed as a little circle going around a fourth space dimension.

be rolled up individually like the garden hose. But they could also be combined together and folded into a tiny sphere. Mathematicians refer to these alternative shapes as different *topologies*. The number of possible topologies rises sharply as more and more extra dimensions are added. When discussing the shape of space, therefore, we must specify not only how many dimensions there are in total, but which ones are 'large' (like the three dimensions we see) and which are compactified (and so unseen). Furthermore, we must specify the geometry and topology of the compactified dimensions too. Later we shall see that extra compactified dimensions of space play a crucial role in string theory and other attempts to construct a unified theory of physics.

There is a second way in which an extra dimension might be hidden from casual view. What if we are somehow trapped, *Flatland*-like, in the three space dimensions we observe? That is, the particles of our bodies, and all the objects around us, 'live' in a three-dimensional space, and are not free to move in the extra dimension. Obviously this trapping arrangement must also affect light, or we would be able to see the fourth dimension even if we couldn't move in it. This trapping idea isn't completely ad hoc, but emerges naturally in some theories of fundamental physics known as 'brane' theories, by analogy with the word 'membrane' (being a two-dimensional sheet embedded in three-dimensional space). Thus, our three-dimensional universe could be a three-brane embedded in four space dimensions. Brane theories are popular in some quarters, and they make specific predictions about the nature of gravitation and the behaviour of subatomic particles that might be testable. As yet, though, there is no experimental evidence that we are living in a three-brane, just a lot of intriguing mathematical theory.[18]

The fact that space could have any number of 'large' dimensions raises the fascinating question of why nature has selected three. Is there anything special about the number three? In the 1950s the English mathematician Gerald Whitrow[19] pointed out that if space had four dimensions and the laws of gravitation and electromagnetism remained unchanged, we'd be in for trouble. The inverse square law would become an inverse cube law, and a simple investigation reveals that planetary orbits would be unstable: Earth would soon spiral into the sun. Similar stability problems afflict atoms. In five or more dimensions the problems get worse. And if there were just two dimensions, waves would have trouble propagating and reflecting, leading to complicated effects that might compromise the ability of complex systems to behave coherently. Whitrow concluded that life would be impossible in a space with a dimensionality other than three. Two dimensions are too few, four dimensions are too many, but three dimensions, Goldilocks-like, are just right. So we have come across another one of those bio-friendly features in need of explanation.

Key points

- The universe began with a hot big bang 13.7 billion years ago, and is still expanding. The expansion is best envisaged as the stretching of space between the galaxies.
- Space is filled with heat radiation that is a remnant, or afterglow, of the hot early phase of the universe. The cosmic microwave background (CMB) radiation has been intensively studied, most notably by the WMAP satellite, because it contains lots of important data about the history and structure of the universe.
- Fluctuations in the CMB reveal the seeds of the large-scale structure of the universe (clusters of galaxies).
- The universe has no discernible centre or edge.
- Because light travels at a finite speed, there is a maximum distance, or horizon, in space beyond which we cannot see.
- Gravitation is described by Einstein's general theory of relativity in terms of warped space (strictly, warped spacetime). Spacewarps are familiar to astronomers.
- Even though space is warped locally, by stars and galaxies, overall the geometry of the universe seems to be flat (Euclidean). Einstein's general theory of relativity then predicts that the universe should have zero total mass: the positive mass-energy of matter is exactly cancelled by the negative mass-energy of the gravitational field of all the matter in the universe.
- There may be additional dimensions of space over and above the three we perceive. Some theories of physics require this. Extra dimensions can be concealed from view, for example by rolling them up to a tiny size.

3

How the Universe Began

Relics of the early universe

I remember attending an astronomy course at London University in 1967 when the lecturer alluded to the then newly discovered cosmic microwave background radiation. He told us that, based on what could be deduced from this discovery about the hot, dense early state of the universe, it was possible to work out the physical changes that took place in the first few minutes following the big bang. Everybody in the audience burst out laughing. In those days it was hard to take seriously a detailed analysis of the universe as it was mere minutes after its origin. Even the Biblical account is limited to 'the first day'. Yet within a few short years the 'early universe' – defined by cosmologists as lasting from about one microsecond after the big bang up to the epoch of WMAP's images (380,000 years) – had become a routine subject for lecture courses and PhD projects. Steven Weinberg's bestseller *The First Three Minutes* was already considered an orthodox review when it was published in 1977.

Let me start by mentioning some straightforward achievements of the hot big bang theory. The biggest success has direct relevance to the central theme of this book: the bio-friendliness of the universe. In Chapter 1, I listed some of the prerequisites for life. Top of the list is a supply of the various chemical elements used by living organisms. These elements did not exist at the moment of the big bang; it was far too hot. After decades of research, it is now possible to reconstruct in some detail how the elements were produced. The story begins about one second after the big bang. At that time, the temperature would have been a searing ten billion degrees, which is hotter than the hottest star. Under these conditions, atoms could not survive. Even atomic nuclei would be smashed apart. So the state of the universe at one second must have consisted of a soup (more accurately, a plasma) of freely moving atomic components – protons, neutrons and electrons.

The simplest chemical element is hydrogen, the nucleus of which consists of a single proton. Most of the protons that came out of the big bang remained free, and were destined to form hydrogen atoms once the universe had cooled enough for each proton to capture an electron. (That final step didn't happen for nearly 400,000 years.) Meanwhile, however, not all the protons were left isolated. Some of them collided with neutrons and stuck to form deuterium, a relatively rare isotope of hydrogen with one proton and one neutron apiece in each nucleus. Other protons became incorporated into helium, the next simplest element, which has a nucleus consisting of two protons and two neutrons. What I am describing is called nuclear fusion, a process which is very well understood. Protons and neutrons could begin combining together to make composite nuclei only once the temperature had fallen enough so that the newly minted nuclei would not immediately be fragmented again by the intense heat. The window of opportunity for nuclear fusion was limited, however, opening up at 100 seconds or so and closing again after only a few minutes. Once the temperature dropped below about a hundred million degrees, fusion ground to a halt because the protons lacked the energy to overcome their mutual electrical repulsion.

It's possible to calculate how much helium was made and how many protons were left over to form hydrogen, assuming that the basic idea of a hot big bang is correct. The answer comes out at about three hydrogen atoms for every helium atom, and virtually nothing else (except for tiny amounts of deuterium and lithium). This is very much what astronomers have measured for the relative abundance of the simplest elements. How do they know? All chemical elements have, imprinted on the light they emit, a distinctive 'barcode' in the form of lines visible in their spectrum, and by analysing starlight astronomers can read the barcodes and tell what the star is made of. The same technique will work for any astronomical source (or absorber) of light, including diffuse clouds of gas. From such measurements we know that the universe is made almost entirely of hydrogen and helium in a roughly three-to-one ratio. Helium, then, is a relic of the first few minutes of the universe.[1]

I mentioned that nuclear fusion is well understood. In fact, pretty much all of the basic physics pertaining to the time between just a millionth of a second and several minutes after the big bang is now regarded as routine. And I don't mean just in theory: the physics of the early universe can be directly tested in the laboratory. On Long Island near New York a large machine called a heavy ion collider is designed to slam the nuclei of gold and other heavy atoms together head on, with enough force to recreate the conditions of the early universe as they were one-millionth of a second after the beginning, when the temperature was over a trillion degrees. These high-energy encounters enable physicists at the Brookhaven National Laboratory, which operates the collider, to experience at first hand what happened when the universe we observe today was squeezed into a volume of space no larger than the solar system, with a temperature almost a million times hotter than the centre of the sun. It turns out that under these extreme conditions even protons and neutrons cannot exist as discrete entities. Instead, they were melded into an amorphous cocktail of subnuclear fragments.

The big bang quickly ran out of oomph

The processes I have been discussing took place in a universe that was expanding and cooling extremely rapidly. By way of illustration, the size of the observable universe roughly doubled between one and two microseconds (a millionth of a second) after the big bang. This represents a rate of expansion some trillion trillion times faster than today. But this frenetic pace didn't last long: by one second the expansion rate had plunged to a trillionth of what it was at one microsecond. The reason for the large deceleration is easy to understand. Gravitation, the universal attraction between all forms of matter, served as a brake on the expansion, all the more so because of the extraordinarily compressed state of matter at the time. The slowing effect continued over the subsequent minutes, hours, years and millennia, but the *rate* of slowing – the magnitude of the deceleration – diminished as the

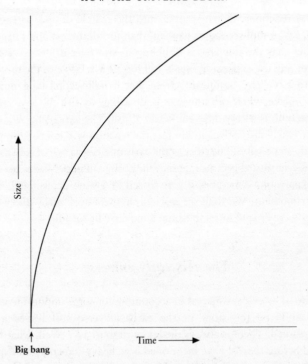

12. *Gravity slows the expanding universe*

How the size of the universe should increase with time according to the general theory of relativity. It starts out expanding explosively fast at the big bang origin, but progressively slows as the attractive force of gravitation acts like a brake.

universe expanded and its gravitational grip weakened. The general trend is shown in Figure 12. Later I shall explain how the slowing effect may have reversed, and turned into a quickening expansion, after a few billion years had elapsed. But for now we are concerned only with the early universe, when the expansion rate was fiercely decelerating.

This basic trend of rapid but decelerating expansion touches on an important issue, which is how the rate of expansion compares with the pace of change of the physical processes taking place, for example nuclear reactions. If, ten seconds into the cosmic drama, a reaction takes, say, a millisecond to happen, then for all intents and purposes the fact that the universe is doubling in size over a few seconds is immaterial: the expansion rate is still too slow relative to the reaction rate to have any significant effect. But a millisecond-long process taking place when the universe is doubling in size in, say, *half* a millisecond, is a very different kettle of fish. The expansion will now be extremely disruptive, and in fact the reaction will fail to keep pace with the expansion. The story of the early universe is one of see-sawing fortunes in this respect. Sometimes the cosmic material keeps up with the expansion, sometimes it gets frozen in a state determined at an earlier moment. We shall see in later chapters how such 'freeze-out' events were crucial in establishing a universe fit for life.

The very *early universe*

In spite of its evident success in accounting for some important astronomical facts, the story of the early universe still leaves a lot unexplained. There is always the temptation to ask, 'Well, what happened before *that*?' I have mentioned how heavy-ion collision experiments can simulate the state of the universe as it was about a microsecond after the big bang. Can we push back even further, to probe still higher energies and reach even nearer to the enigmatic first moment when the big bang went bang? The world's largest particle accelerators can attain energies much greater than Brookhaven's machine. Although they are normally restricted to two-particle collisions, these accelerators nevertheless enable physicists to glimpse some of the processes that would have been going on in the universe a mere *trillionth* of a second after the big bang.

The urge to slice up the time-line of creation into ever finer subdivisions is dictated not so much by cosmologists' obsession with detail as by the remorseless arithmetic of scaling. As the great cosmic movie is played in reverse back to the moment of origin, so the pace of change escalates. About as much happens in the interval between a microsecond and a millisecond as happens between a millisecond and a second, or between one second and several minutes. The reason is that both the compression and the temperature of the universe rise without limit as the crucial moment of time zero is approached. This, combined with the fact that the deepest structures of the universe were forged at the earliest times, inevitably drives the compulsion to keep asking, 'So what happened before?'

Flushed with the success of their early-universe theory, cosmologists in the 1970s began to take an interest in what they called the *very* early universe. A microsecond may be a short interval by human standards, but in the world of subatomic particle physics it is extremely long. Many observed reactions, such as the decay of certain subnuclear particles, take only about a trillion-trillionth of a second. This brief interval establishes a basic timescale for particle physics, and cosmologists were understandably drawn to speculate on what might have taken place in the universe at, and even before, such an interval of time had elapsed.

This wasn't just idle curiosity. It was clear in the 1970s that some very basic features of the universe remained completely unexplained – indeed, they were positively mysterious. First, and most obvious, was the problem of what actually caused the big bang. A related question was why the big bang was just *that* big, rather than bigger or smaller: what, precisely, determined its oomph? Then there was the puzzle of why the large-scale geometry of the universe is flat, and the related mystery of why the total mass-energy of the universe is indistinguishable from zero. But the biggest puzzle of all concerned the extraordinary uniformity of the universe on a grand scale, as manifested in the smoothness of the CMB radiation. As I have pointed out, on a scale of billions of light years the universe looks pretty much the same everywhere. And similar remarks apply to the expansion: the rate is identical in all directions and, as best we can tell, in all cosmic regions. All these features were completely baffling in the 1970s, yet they are all crucial for creating a universe fit for life. For example, a bigger bang would have dispersed the cosmological gases too swiftly for them to accumulate into galaxies. Conversely, had the bang been not so big, then the universe would have collapsed back on itself before life could get going. Our universe has picked a happy compromise: it expands slowly enough to permit galaxies, stars and planets to form, but not so slowly as to risk rapid collapse.[2]

Why is the universe so smooth?

On the face of it, an explosion is an unlikely way to create a smooth and orchestrated expansion; explosions are normally rather messy affairs. If the big bang had been slightly uneven, so that the expansion rate in one direction outstripped that in another, then over time the universe would have grown more and more lopsided as the faster galaxies receded. We don't see that. Evidently the big bang had exactly the same vigour in all directions, and in all regions of space, tuned to very high precision. In itself, this would be enigmatic enough, but it looks downright contrived when we remember the existence of the horizon. Imagine two regions of space on opposite sides of the sky, A and B, each 10 billion light years from Earth. They look very much the same, containing similar distributions of galaxies with similar red shifts. Located poles apart, these cosmic regions are seen by us today to be separated by about 20 billion light years from each other. Since the universe is less than 14 billion years old, light cannot have had time since the big bang to have travelled from one region to the other. An inhabitant in region A would not be able to see region B, or know of its existence, even though human beings, situated as we are between them, can see both (see Figure 13). The horizon around A will not yet have extended as far as B. Regions A and B evidently cannot know about each other. This means that they are what is termed causally independent, because no object or force can travel faster than light (see Box 1, p. 36), so no physical influence can have linked these regions. What has happened in region A cannot (in this simple picture) possibly affect what has happened in B, and vice versa. Why, then, do A and B look more or less identical? How has the universe con-trived its explosive genesis with such exactitude that there is no dis-tinguishable difference across the sky, even between regions that have never been in causal contact? It's as if a troupe of blind and deaf ballerinas were to perform a perfectly choreographed dance. To throw the problem into stark relief, at the epoch from which the CMB emanates the observed universe contained *millions* of apparently causally independent domains, and yet – as I have stressed – this radiation is astonishingly uniform. How has the entire cosmos

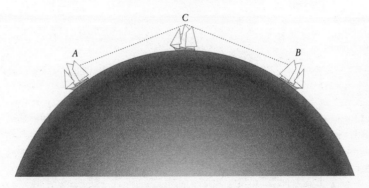

13. *Whose horizon?*
Three boats sail in line astern. A lookout on boat *C* can see boat *A* (ahead) and *B* (astern). But a lookout on *A* cannot see *B*, and vice versa: they lie over each other's horizon. Similarly in the case of the cosmological horizon, we on Earth can see very distant regions of the universe on opposite sides of the sky which are so far apart from each other that light hasn't had time to travel between them since the big bang. These regions should therefore be 'causally disconnected'.

cooperated to achieve this? Was it just an incredible fluke, or did some physical process operate in the very early universe to bring about such a very special state of affairs?

The inflation theory explains it all at a stroke

The germ of an answer was found in the early 1980s by Alan Guth, a theoretical physicist now at the Massachusetts Institute of Technology. It is based on an extremely simple idea called *inflation* (to distinguish it from the normal cosmic expansion). In Guth's original version, inflation worked like this. First there was the traditional big bang, which didn't need to be uniform or orchestrated: a messy bang would do, just to get the cosmic show going. Then, a split second later, the universe suddenly jumped in size by a truly enormous factor (see Figure 14). How enormous? Guth suggested at least 10^{25} (ten trillion trillion), implying that the entire observable universe leapt

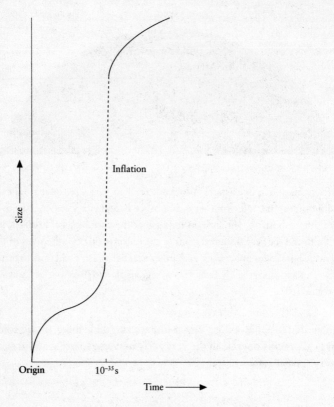

14. *Inflation*

In Alan Guth's original inflation theory, the very early universe goes through a short burst of enormously accelerating expansion, leaping in size by a colossal factor in a split second (the graph is not drawn to scale). When inflation stops, normal decelerating expansion is resumed, matching the form shown in Figure 12. In this simplified picture the universe is shown originating with a conventional big bang followed by a brief period of deceleration prior to the inflationary phase. However, inflation erases the traces of the prior phase, so this part of the curve should not be taken too seriously.

from about the size of a proton to the size of a grapefruit virtually instantaneously. The actual magnification factor was unimportant so long as it was very big. Finally, inflation shuddered to a halt and normal expansion resumed, dovetailing into the orthodox story of the early universe I have already related.[3]

Shortly I'll come to what might have caused the universe to behave in this remarkable fashion, but first let me explain how inflation neatly solves the problem of cosmic uniformity. Any initial irregularities would be 'stretched to death', in the same way that an inflating balloon loses its crinkles. In effect, inflation obliterates all records of previous complexity and generates a naturally smooth universe. And there is a bonus. Just as inflating a balloon reduces the curvature of its surface, so inflating space makes it less and less curved. Inflate it enough (and a factor of 10^{25} is enough) and it will be indistinguishable from flat. So inflation explains both the uniformity and the flat geometry of the universe.

Antigravity

The way I have described it, Guth's inflation seems little more than a magic wand. It would have fallen on deaf ears had Guth not provided a credible physical mechanism to explain how inflation might have occurred. Luckily he already had such a mechanism at hand: it involved modifying the normal role of gravitation in cosmology. The gravitational pull of the universe serves to diminish the expansion rate progressively. Inflation does just the opposite: it is a brief episode in which the expansion rate accelerates hugely, causing the universe to swell up super-fast. Guth proposed that a type of antigravity force was responsible. This isn't quite as fanciful as it sounds: it so happens that antigravity is built into Einstein's general theory of relativity in a very natural way (see Box 3). In fact, antigravity turns out to be the default option in the theory of relativity.

Imagine a region of empty space, pure and simple. According to general relativity, the natural behaviour is for this region to expand faster and faster. (There is another, slightly less natural, possibility

3. *Einstein's greatest mistake*

Einstein proposed his model of a finite but unbounded universe in 1917, before Edwin Hubble announced that the universe is expanding. Einstein therefore assumed that the universe is static. This presented a problem, because it looked as if a static universe would collapse under its own weight. What would shore it up against the universal gravitational pull between all the bodies? Einstein had an ingenious solution. He proposed that the normal force of attractive gravity is counterbalanced by some sort of antigravity, or cosmic repulsion. He looked to the equations of his general theory of relativity for a clue, and found that if he added an extra term to his 1915 equations, then the theory would describe *two* forms of gravitation – the normal attractive force, and a repulsive, or antigravity, force. (The antigravity can be thought of as due to the energy of empty space – see p. 68.)

The antigravity force has the unusual property that it *increases* with distance, making it quite unlike the normal attractive force of gravitation. But it was precisely this unusual property that Einstein could exploit. He reasoned that the repulsive force must be very weak on the scale of the solar system, or it would upset the impressive agreement between his original equations and astronomical observations of, for example, the motion of Mercury's orbit (see Box 2 on p. 43). But on an extragalactic scale the repulsion would become significant, and could rival the force of attraction. General relativity gave no clue about the overall strength of the repulsive force (physicists call it a 'free parameter'), so Einstein was able to propose a value that would exactly counterbalance the weight of the universe, thus preventing its collapse.

That was the situation in 1917, when Einstein published his modified equations, including the extra term. He was able to find a solution of them which corresponded to a static universe, resting in equilibrium between attractive and repulsive aspects of gravitation. That solution is the hyperspherical model described in Chapter 2 (see p. 44). In 1930 Einstein travelled to the United States and met Hubble, and learned about the observations indicating an expanding universe. Einstein immediately recognized what a mistake it had been to preoccupy himself with a model of a static universe. Had he stuck to his guns, and worked with the original equations – without the extra antigravity term – he would have been forced to conclude that the universe must be either expanding or contracting, and he would surely have opted for the former.

In fact a Russian scientist, Alexander Friedmann, solved Einstein's original equations in 1921 and thus derived a variety of expanding and contracting cosmological models. He sent the solutions to Einstein for his consideration, but the great man was lukewarm about them, obviously preferring his own, static, model. As a result of this blinkered thinking, Einstein missed the chance to predict one of the great observational discoveries of twentieth-century science, which would have provided a further triumphant test of his cherished theory of relativity. Having eventually realized his mistake, Einstein abandoned the antigravity term in disgust, calling it 'the biggest blunder' of his life. As a result, antigravity went seriously out of fashion for decades. When I was a student in the 1960s very few cosmologists were prepared to speak for it.

History, however, has a habit of springing strange surprises. To be sure, following Hubble, antigravity was no longer needed to describe a static universe. But that does not logically mean that the force doesn't exist. In the late 1990s astronomers announced that the expansion rate of the universe is picking up, an effect they attribute to a universal antigravity force, currently indistinguishable from what Einstein proposed in 1917. So Einstein's greatest mistake may yet turn out to be a triumph!

that it would shrink.) The only exception would be if the energy of that region of empty space is *precisely zero*; then it will remain inert. Now you might imagine that if space is empty it should indeed have precisely zero energy – because there is nothing in it! Which is true, in a sense. However, that ignores the possibility that invisible fields may permeate space. Such fields would contain energy. The gravitational effect of field energy depends on the nature of the field: some fields (e.g. electric) would make the universe shrink; others would create antigravity and make it expand. The latter include so-called *scalar fields*, and these were the ones that interested Guth.[4] To be honest, nobody has yet observed a scalar field, but physicists have good theoretical reasons for supposing that they exist. Experimenters are hot on the trail of one such scalar field, called the Higgs field, which I shall discuss in Chapter 4. Guth wasn't too concerned about the experimental evidence for scalar fields. He just assumed that there is a suitable one out there that will do the inflation job, and he called this hypothetical entity the *inflaton field*.

I haven't explained why the inflaton field antigravitates. The reasons are a bit technical, but I can convey the gist. In Newton's theory, gravitation is generated by mass. In Einstein's general theory of relativity, mass is also a source of gravitation, as is energy (remember that Einstein's equation $E = mc^2$ tells us that energy has mass). But it doesn't stop there. Pressure too is a source of gravitation in the general theory of relativity. We don't normally think of pressure creating a gravitational field, but that is because its effect is tiny in most circumstances. Even the immense pressure inside Earth, for example, contributes less than a microgram to your body weight.[5] But if the pressure gets seriously big, it can rival the energy in its gravitating power. 'Seriously big' here means the sort of pressure found inside a collapsing star rather than a planet. Another example, however, is a scalar field: it has a pressure comparable to its energy.[6]

But why does the scalar field produce *anti*gravity? The crucial factor is the pressure: for a scalar field it is negative. Negative pressure isn't especially exotic: it is no more than what we normally call tension – a stretched elastic band provides a familiar example. In three dimensions, a block of rubber pulled in all directions would have negative pressure. Now negative pressure implies negative gravitation – a repulsive, antigravity force. So a scalar field generates gravity by virtue of its energy, but antigravity by virtue of its (negative) pressure. A calculation shows that the antigravity beats the gravity by a factor of three, so the net effect of the scalar field is to antigravitate.[7]

The origin of matter

To return to the story of inflation, Guth proposed that during the first tiny fraction of a second after the birth of the universe a scalar field permeated space, exerting an immensely strong antigravity effect and causing the universe to embark on a phase of runaway expansion, getting faster and faster all the time. Having identified that a scalar field would do the trick, his next task was to determine whether the antigravity effect would be strong enough to overpower the colossal gravity produced by all of the normal matter in the universe. He took his cue from the subject of grand unified theories, or GUTs – attempts to meld together three of nature's fundamental forces. (This is a subject I shall explore more deeply in Chapter 4; suffice it to say here that scalar fields play a central role in GUTs.) Guth assumed that his inflaton field would be one of the GUT-type scalar fields, which gave him an all-important quantity: the strength of the field. Plugging its value into his inflationary calculations, he discovered that the antigravity would not only easily overwhelm the universe, it would be so strong that the universe would double in size every 10^{-34} s (i.e. every hundred trillion-trillion-trillionths of a second). Nothing in nature has been observed to happen so fast. To put it in perspective, in the time that a patch of the universe doubles its radius, light would travel a mere trillion-trillionth of a centimetre – nowhere near enough to cross even an atomic nucleus. This was rapid expansion indeed.

Inflation is a very attractive idea, and most cosmologists are sold on it. However, a crucial issue is how it came to an end. How would the universe have extricated itself from stupendously rapid runaway expansion? Guth suggested that the inflaton field was inherently unstable, and was thus condemned to a fleeting existence. He proposed that it decayed away after only about 10^{-32} s, following which the universe would resume its normal, decelerating expansion. This duration doesn't seem very long, but such is the rate of inflation that in 10^{-32} s the universe would have ballooned out by a huge factor. Any matter present before inflation would have been diluted to a negligible density, leaving the universe effectively empty – a vacuum. Obviously a vacuum isn't a good description of the universe today, or even at one second. Where, then, did all the matter – the electrons, protons, neutrons, and so on – come from, once inflation had ceased?

The theory has a ready answer. The enormous energy stored in the inflaton field during the inflationary phase must go somewhere when the field decays, and that somewhere is into heat. The cosmic microwave background pervading the universe today – the afterglow of creation – represents the remnants of the inflaton field energy. In effect, the dizzying energy of expansion during the inflation phase is converted into the heat energy of the big bang, energy which now bathes the universe. The next step is to turn heat into matter. Einstein's formula $E = mc^2$ tells us that so long as there is enough energy E to pay for the mass m of a particle, the way lies open for matter to be created. Putting into the formula a value for the speed of light c, and translating from energy into temperature, we find that a billion degrees – the temperature of the universe at about one second – is hot enough for heat energy to create electrons. At earlier times, when the temperature was higher still, heavier particles such as protons would have been created. At the end of inflation, the intense energy released would have heated the universe to around a thousand trillion trillion degrees – more than sufficient to create all 10^{50} tonnes of matter in the observable universe.[8]

The problem of exiting gracefully from inflation

Inflation was a dream theory, the like of which rarely appears in science. In one fell swoop it solved several puzzles about the structure of the universe, and it made very specific predictions which could be tested. The foremost of these – that the universe should be geometrically flat – was made at a time when the astronomical evidence was suggesting otherwise. But then a series of observations, culminating in the results from WMAP, confirmed the flatness prediction, providing powerful support for the inflation theory.

Once the general idea of inflation had entered cosmology, it was there to stay. Guth's original theory, however, contained a fatal flaw – the so-called graceful exit problem. The decay of the inflaton field is a quantum process, so its initiation is subject to the usual unpredictable quantum fluctuations. As a result, it would decay at different times in different places, in the form of randomly distributed bubbles – bubbles of space, that is, in which the inflaton field had decayed, surrounded by regions of space where it had not. The energy given up by the decayed inflaton field would be concentrated in the bubble walls. Bubble collisions would release this energy, as heat, but the process would be utterly chaotic and generate as much inhomogeneity as inflation was designed to remove. These shortcomings were addressed by a number of distinguished cosmologists who found the idea of inflation compelling.[9] The solution was to find a theoretical scheme that would avoid bubble collisions and enable the bubbles to grow to a size much larger than the observable universe. One way to do this, called eternal inflation, has an important bearing on the issue of why the universe is bio-friendly, and I shall explain it in some detail later.

Ripples from the edge of time

If inflation stretched space by a truly enormous factor, you might expect the universe to be exceedingly smooth at the end of it. But if it had been, there would be no prospect for life: without galaxies and stars, life would be impossible. The hot and cold fluctuations in the CMB shown in Figure 3 (p. 21) are the seeds of large-scale structure. But where did these fluctuations come from? How did inflation produce a universe which is almost, but not quite perfectly, smooth?

A compelling answer – although it may not be the right one – was already at hand at the time of the original inflation theory. In fact, it predated it. The cause of the cosmic ripples, it seems, might very well lie with quantum mechanics. Readers familiar with the basic ideas of quantum mechanics will know that Heisenberg's uncertainty principle leads to irreducible fluctuations in all physical quantities (those readers who are not familiar with quantum mechanics will find a bare-bones summary in Box 4). By applying quantum mechanics to inflation we can predict that some regions of space inflate slightly more or slightly less than others, thus producing a frothy structure superimposed on the overall smoothness of the universe. Conspicuous

4. *Quantum weirdness*

Twentieth-century physics is built upon two revolutionary theories: relativity and quantum mechanics. Both form an essential part of our description of the universe. Quantum mechanics began as a theory of matter on the atomic and subatomic scales, but most physicists believe that it applies to everything, including space and time, on all scales. Nevertheless, conspicuous quantum effects are confined mainly to the microscopic scale. (A notable exception might be the large-scale structure of the universe – which is very conspicuous!)

The quantum theory began in 1900 with the suggestion by Max Planck that when heat radiation is emitted by a hot body, it comes out only in discrete little packets, or *quanta*. Einstein extended this idea to photons of light, which he treated as if they were tiny particles. However, electromagnetic radiation such as heat and light is also known to behave as a wave, so these early ideas of quanta implied, bizarrely, that light could have both a wave and a particle aspect, which caused much confusion. In the 1920s it was found that particles of matter such as electrons could also display wave-like properties. It became clear that the nature of reality in the atomic realm was very weird, and that *wave–particle duality* was a basic feature. Which aspect – wave or particle – manifests itself depends on the type of experiment or observation performed. It is not possible to say in general whether a photon or an electron (or a proton, or neutron, . . .) is 'really' a wave or a particle, because it can behave like both.

Closely related to this vagueness is a central tenet of quantum mechanics called *Heisenberg's uncertainty principle*. This forbids a quantum object from possessing a full set of familiar physical attributes at any given time. In daily life we might ascribe to an object such as a ball several properties, for example position, speed, spin and energy. These are properties also possessed by subatomic particles such as electrons, *but they do not all take on precise values at once*. We might locate an electron at a point in space, say, thus attributing a well-defined position to it, but then, according to Heisenberg's uncertainty principle, we cannot also assign to it a well-defined motion. Alternatively we might determine the electron's velocity, but then its position would be ill defined. The degree of this quantum fuzziness is not arbitrary, but is precisely specified by Heisenberg's uncertainty principle. The numbers are such that quantum uncertainty is important for atoms and subatomic particles, but becomes far less important for progressively more massive or complex systems. The scale of quantum effects is fixed by a parameter called *Planck's constant*, the value of which can be determined experimentally. It is denoted by the letter h. It is one of a number of fundamental constants of physics, alongside the speed of light c and Newton's gravitational constant G (see Box 6, p. 88).

Most physicists interpret quantum uncertainty to be intrinsic, and not merely the result of human ignorance or clumsiness in measurement. One might express this by saying that even the electron is uncertain of its properties. Thus quantum uncertainty cannot be ameliorated by 'looking harder'. In this respect, it stands in contrast to random chance in, say, a game of roulette, or movements in the stock market. Changing share prices certainly have their causes in something: if they appear random and unpredictable it is because humans do not have all the information they need to work out how they vary. Quantum randomness, by contrast, is irreducible, which is to say that quantum processes are in some sense genuinely spontaneous – without any specific cause.

Quantum uncertainty can sometimes be envisaged, very roughly, in terms of *fluctuations*. One can think of a property, such as the position of an electron, as fluctuating: the electron jiggles about spontaneously in an unpredictable manner. All measurable quantities are subject to quantum fluctuations, to an extent prescribed by Heisenberg's uncertainty principle. One consequence is that two identical situations might produce different outcomes. For instance, imagine firing an electron directly at an atom: it may deflect to the left or to the right with equal probability. If you do the experiment today, perhaps it deflects to the left. But you could do the experiment tomorrow, *under identical circumstances*, and the electron may on that occasion deflect to the right.

Although the theory of quantum mechanics can provide the betting odds, generally one cannot know in advance what will happen in any specific case. A familiar example is radioactivity. Uranium nuclei are unstable and decay over a period of billions of years. Each nucleus has a certain probability of decaying during a fixed time interval, via a random quantum process, but it is impossible to know in advance when a *given* nucleus will decay. Similarly for the decay of an excited atom emitting a photon: the process has a definite probability, but a given event cannot be reliably predicted. Quantum uncertainty applies not just to particles, but to fields too, so that, for example, an electromagnetic field is subject to random variations in strength, even in a perfect vacuum in which the *average* field strength is zero. These so-called vacuum fluctuations can be given an equivalent description in terms of 'virtual photons' that spontaneously appear and then disappear in empty space (see p. 111 for a discussion of virtual particles). Vacuum fluctuations turn out to be crucial to our understanding of dark energy and antigravity, on which the fate of the entire universe hinges. The inherent indeterminism in nature is what led Einstein – who hated quantum mechanics – to declare (incorrectly) that 'God does not play dice with the universe.'*

Quantum mechanics has lots of other weird features. For example, it predicts that particles can possess an intrinsic spin which must have a fixed number of basic units, and can point only in certain prescribed directions. Quantum particles can tunnel through force barriers, swerve round corners or exist in many places at once. Some of these properties will turn out to be important for the story of the universe, especially the very early universe.

* *The Born–Einstein Letters*, translated by Irene Born (Macmillan, London, 1971), p. 91.

quantum fluctuations are normally restricted to the atomic scale. If this explanation for the cosmic ripples is correct, atomic-scale fluctuations were hugely inflated and writ large on the sky.

I actually did some work on this topic myself. In the 1970s I was in the Mathematics Department of London University's King's College, trying to understand quantum effects in various cosmological settings. There was a general feeling that although quantum mechanics would be irrelevant to the dynamics of the universe today, it must have been important near the origin, when the universe was very compressed. I was assisted in my work by a student named Tim Bunch. Tim and I decided to look at quantum effects in a universe that expanded at an exponential rate – that is, it keeps on doubling in size in successive fixed intervals of time. We chose this particular model universe, known to astronomers as de Sitter space after its original description by Willem de Sitter in 1917, not because we thought the real universe resembled it, but because by using it we could solve the equations exactly. In theoretical physics, one exact solution is worth a hundred numerical approximations.

So we decided to do quantum theory in de Sitter space. We found that in many respects the expansion of space had no effect. This came as a bit of a surprise at the time, because most calculations showed that the expansion of the universe generally causes particles (or quanta) such as photons to be created in otherwise empty space – from the vacuum.[10] It happens because the expansion disturbs and excites any fields, such as the electromagnetic field, which might be pervading space. The effect is normally extremely small, although it may have been important just after the big bang. Anyway, we found that in de Sitter space there was *no* such particle production, a curious result that can be traced back to the exponential nature of the expansion, and the underlying symmetries of spacetime that this implies. But that is not to say that the expansion of space in de Sitter's model has no quantum effect at all. It does. In particular, the vacuum state of de Sitter space was still subject to quantum fluctuations, which may loosely be regarded as particles being created but then rapidly destroyed again, coming and going in an ephemeral dance (these are so-called virtual particles – see p. 111). There is no net gain or loss of particles, then, but lots of fleeting quantum activity.

When we were doing this work in the late 1970s, we had no idea that within a few years it would turn out to be just what was needed to describe the density fluctuations in the inflationary universe. By good fortune, the model we had picked – de Sitter space – turns out to be an exact description of how the universe behaves when it is inflating. We had chosen de Sitter space for rather more mundane reasons: to get Tim a PhD without having to use a computer to finish the calculation! Such is the march of science.

What happened before the big bang?

Most people are prepared to accept that the universe as we know it began suddenly with a huge explosion, but they inevitably ask two related but difficult questions: What caused the big bang? What came before it? Questions about the origins of things are endlessly fascinating, and none more so than the ultimate origin of the universe itself. Reflecting on origins has the startling effect of a Zen koan: how can something come to exist that did not exist before? Only one sort of explanation seems intuitively satisfactory: the new entity must arise somehow from the transformation of an earlier, different, entity. 'Nothing can come out of nothing,' wrote Lucretius.[11] 'There is no such thing as a free lunch' might be the modern equivalent. Maybe. But when the entity concerned is the entire universe, this simple dictum may have exceeded its limits. Some cosmologists, at least, believe that the universe may be the ultimate free lunch.

Can inflation theory help us to understand the cause of the big bang? The answer is yes and no. By its very nature, inflation erases the record of what went before. This is a key point. Fred Hoyle, who as I have already mentioned was critical of the entire big bang idea, used to quip that the big bang theory merely says that the universe is the way it is because it was the way it was. In the early days of the theory, that was true. The big bang itself was unexplained: an event postulated to account for the facts, but an event without an apparent cause, and seemingly lying beyond the scope of science altogether. To explain the universe we observe, it was necessary to put into the theory by hand the requisite initial conditions that would lead to what we see today, without any justification. One could equally well put in any initial conditions and obtain a description of any universe one chose. But inflation addresses this concern by enabling us to explain many of the fundamental features of the universe as a product of physical processes during inflation rather than attributing them to ad hoc initial conditions. Good though this may be, there is a downside: it seems to place the *ultimate* origin of the universe out of reach. On the other hand, the same theory that describes inflation can give us clues about how inflation got going in the first place, and may offer pointers to what physical state preceded it.

Let me approach this topic step by step. Discussions about the ultimate origin of the universe are notoriously slippery, so I want to proceed carefully to avoid adding to the confusion. I shall start by ignoring inflation for a moment and adopting an obviously unsound model of the universe: a perfectly round ball of matter surrounded by an infinite void. We know that the universe is expanding, so the ball had better be getting bigger with time. In the past it was smaller. If we run the expansion in reverse for 13.7 billion years, then the ball shrinks to a single point, a single, sizeless dot. And then . . . ? Nothing – the ball has vanished! Play the sequence forwards, and the universe appears from nothing at a single point, balloons out and eventually expands to cosmic proportions. Now, let's consider what is meant by 'nothing' in the foregoing description. Clearly it is empty space. If this account captures the essential manner in which the universe came into existence then we are left with a puzzle. Why should a ball of matter suddenly appear out of nowhere, at some particular moment in time and at some particular location in pre-existing empty space, when this event hasn't happened for all eternity up to that moment? What would cause it to happen, and happen just *then* and just *there*? There is no satisfactory answer.

A similar conundrum afflicted early Christian theology. 'What was God doing before he made the universe?' taunted the disbelievers. If God is perfect and unchanging, as the theologians claimed, then there is nothing to single out a particular moment in time for the creation of the universe from the infinity of prior moments when the same God, in the same state, did not create the universe. The slick answer to the question of what God was doing was, 'Busy making hell for the likes of you!' But the criticism is a very real and deep one, addressing the seemingly contradictory notion of a timeless being acting within time. A shrewd answer was provided by Augustine, who spotted that the problem lay not with the nature of God, but with the nature of *time*.

Creation from nothing

I'll give Augustine's answer shortly, but first let me consider a more realistic model of the big bang. The universe is not a ball of matter surrounded by empty space, as I explained at length in Chapter 2. But the *surface* of the ball might be a good representation of *space itself*. Remember the caveat here: space is three-dimensional, whereas the surface of a ball is two-dimensional. So the ball's surface is an analogy, not an accurate description: the interior of the ball and the space outside it are not part of the physical universe being discussed here. They are simply aids to visualization. Some people give up at this point because they cannot picture in their mind's eye a three-dimensional spherical surface (a hypersphere), but I urge you to stay with me.

Again, let's play the movie in reverse. The surface of the ball shrinks down onto the centre until all points of the surface converge at a single point of space. And then . . . nothing. But in this case 'nothing' is not a surrounding void, because the only space that matters – physical space – is represented by the *surface* of the sphere, and that has totally vanished. So this time the nothing before the big bang really is 'no thing' – neither matter, nor space. *Nothing*.

The real universe consists of more than expanding space, of course: there is matter too. As space is compressed to zero volume, the density of matter becomes infinite, and this is so whether space is infinite or finite – in both cases there is infinite compression of matter to an infinite density. In Einstein's general theory of relativity, on which this entire discussion is based, the density of matter serves to determine (along with the pressure) the curvature or distortion of spacetime. If the theory of relativity is applied uncritically all the way down to the condition of infinite density, it predicts that the spacetime curvature should also become infinite there. Mathematicians call the infinite curvature limit of spacetime a *singularity*. In this picture, then, the big bang emerges from a singularity. The best way to think about singularities is as boundaries or edges of spacetime. In this respect they are not, technically, part of spacetime itself, in the same way that the edge of this page is strictly not part of the page.

So the first moment of the universe – in this highly simplified picture – is not a moment or a place at all, but a *boundary* to moments and places. This may seem pedantic, but an important feature of a boundary is that it signals a 'no further!' warning. A boundary to spacetime says that spacetime cannot be continued through it. That is as expected. When a physical theory contains an infinite quantity, the equations break down and we cannot continue to apply the theory. So the big bang singularity is a boundary where the general theory of relativity says, in effect, 'Infinity? Ugh! I give up!' and space and time come to an end. Spacetime singularities, I should say, are not an obscure technicality. Both Roger Penrose and Stephen Hawking made their names in theoretical physics by proving several important singularity theorems in the 1960s using elegant mathematical techniques. Some of my colleagues devoted their entire careers to studying singularities, and the subject even made it into an early episode of *Doctor Who*.

I said that spacetime cannot be continued 'through' a singularity. Strictly speaking, there is no reason why spacetime cannot exist on the far side of a singularity. That is, we can imagine joining another spacetime onto the big bang singularity from the other side. However, that would be entirely gratuitous. Because the singularity represents infinite curvature and density, and an end to the basic physical theory that describes all this, we cannot suppose that any physical object or influence can penetrate the singularity, so there is no way of knowing whether there is anything on the far side of it or not. Nor can we attribute much meaning to the assertion that there is. After all, space and time there would not be 'our' space and time, so even proclaiming that the 'other' spacetime comes 'before' our big bang is moot. If appending this 'prior spacetime' carries no physical consequences for our universe, no purpose is served in positing it.

The big bang as the origin of time itself

The foregoing point touches on another common misconception. I described the singularity in the movie-in-reverse account as 'the vanishing point' of the universe. But why did it have to vanish? Could the singularity not have just sat there? In forward-time description, there would be a singularity – think of a point of infinite density if you like, a structureless, sizeless cosmic egg – existing for all eternity, when suddenly it went 'bang!' In that case, what came before the big bang would no longer be 'nothing', it would be 'a singularity'. Some popular accounts of the origin of the universe promulgate this dubious notion. However, it won't do. The theory of relativity links space and time together to form a unified spacetime. You can't have time without space, or space without time, so if space cannot be continued back through the big bang singularity, then neither can time. This conclusion carries a momentous implication. If the universe was bounded by a past singularity, then the big bang was not just the origin of space, but *the origin of time too*. To repeat: *time itself began with the big bang*. This neatly disposes of the awkward question of what happened before the big bang. If there was no time before the big bang then the question is meaningless. In the same way, speculation about what *caused* the big bang is also out of place, because causes normally precede effects. If there was no time (or place) before the big bang for a causative agency to exist, then we can attribute no *physical* cause to the big bang.[12]

People often feel tricked when they are told this, and sometimes get quite emotional about it, as if the entire discussion is a sly word-game used by wily scientists to flummox their detractors. Cynics say that cosmologists avoid giving a straight answer about what happened before the big bang because they don't know and don't want to admit it. It is true that cosmologists don't know the answer, but that is not because they are unable to come up with some possibilities. The critic's argument usually goes as follows. How can time just switch on like that? *Something* must have preceded the big bang. And it's true that we find it hard to imagine tracing the history of the universe further and further back to a point at which time just stops. But in fact the notion is neither absurd, nor new. Augustine was already there in the fifth century. His considered answer to what God was doing before creating the universe was that 'the world was made with time and not in time'.[13] Augustine's God is a being who transcends time, a being located outside time altogether, and responsible for creating time as well as space and matter. Thus Augustine skilfully avoided the problem of why the creation happened at that moment rather than some other, earlier, moment. *There were no earlier moments.* Identical reasoning applies to the scientific problem. If the universe originated 'in time', then it cannot have been caused by any physical process that has a finite probability, because if it did then the event would already have happened, an infinite time ago. On the other hand, if the universe was made 'with time' then this problem goes away.

I am sometimes asked whether Augustine's prescient comment implied that he had a divine revelation about the cosmic creation. Well, had he written down Einstein's equations rather than penned a dramatic phrase I could believe it. In fact, he wasn't even the first person to hit on the idea of time coming into being with the universe. Plato said much the same thing hundreds of years earlier. The history of philosophy is so rich and diverse that it would be astonishing if theories emerging from science hadn't been foreshadowed in some vague way by somebody. The significant thing about Einstein's work is that he showed in a precise, testable way, using detailed mathematical theory, how space and time are *part of* physics, and not merely a given arena in which the great drama of nature is acted out. It follows that if we are trying to explain the origin of the physical universe, we have no choice but to attempt to explain the origin of space and time too. So the claim that time begins with the big bang is clearly the right starting point.

Was the big bang really a big bounce?

A more serious objection to the account I have given so far is that I have assumed perfectly regular geometry and a universe filled with matter at a uniform density. These are obviously gross idealizations. Imagine a wonky sphere being shrunk without limit. This time the different points on the surface will not converge in a neat and orderly manner to a single point, unless they converge at different rates and conspire to reach the centre at the same moment. Is this likely? No, as it turns out. A disorderly aggregation of particles moving according to the equations of general relativity will generally not all converge to a point, but will mostly miss one another. What happens, then, if you play the cosmic movie in reverse, watch the bits fail to come together, and carry on playing? What you find is that the converging components enter a state of pandemonium and then begin to diverge again. Playing the movie in the forward direction from the remote past depicts a universe that contracts from a large size, collapses violently to a very high density, and surges out again. The big bang has been replaced by a big bounce.

Could the real universe be like this? Logically, there is no reason why not, but there are some serious scientific problems with the idea. The first is that it replaces one problem – Why was there a big bang? – with another: Why was there a contracting universe, with all the right bits in the right places, moving together in the right way to come together in a dense cluster and mimic a coherent big bang? How did this contracting universe come to exist in the first place? Replying that 'it always existed' is hardly an answer. We don't explain something by saying that it has always been there. A variant on this theme is the many-bounces or cyclic universe, in which space expands from a big bang, reaches a maximum point, contracts again to a big bounce, which triggers another cycle of expansion and contraction, and so on, ad infinitum. Again, the existence of such a universe isn't explained merely by saying that it has always been bouncing merrily along its way.

Another objection concerns the so-called second law of thermodynamics. In its most comprehensive statement, this law focuses on irreversible processes – anything that can go in one direction and not the other. A good example is the collapse of a star to a black hole: you can't get the star back out again. Any irreversible processes in the universe which proceed at a finite rate (e.g. star burning, stellar collapse) will reach their final state in a finite time. And in that case, if the universe is infinitely old, then it ought to be in its final state by now. Just like a clock running down, the Great Cosmic Clock should already have stopped ticking. Obviously it hasn't (see Box 5).[14]

5. Why the universe cannot have existed for ever

By the 1850s, physicists knew about the second law of thermo-dynamics, which forbids perpetual motion machines. No engine, for example, can run indefinitely without being resupplied with fuel. For the sun and other stars, the second law of thermo-dynamics spells eventual doom. The sun, which sustains most life on Earth, has been shining steadily (actually getting a bit brighter) for $4^1/_2$ billion years. Today we know that it derives its energy from nuclear reactions in its core; nobody knew that in the 1850s, but obviously there had to be an energy source of some sort, and no energy supply is inexhaustible. The sun can't go on burning for ever: sooner or later its stock of fuel will run out. A back-of-the-envelope calculation reveals that it is about halfway through its life cycle. In another 4 or 5 billion years it will be in trouble, and will end up collapsing to a so-called white dwarf.

The story is similar for other stars: they are not immortal. Stars are born, and stars die. As there is a limited amount of raw material (mainly hydrogen gas) in our galaxy and others, the time will come when no more stars get made, and the existing ones will all be snuffed out, ending their days as black holes, neutron stars or black dwarfs. This general trend was obvious in the nineteenth century, and was referred to as the heat death of the universe. Yet at the time nobody seems to have drawn the obvious conclusion: that the universe cannot have remained unchanging for ever, at least in anything like its present form, or it would already be a stellar graveyard. This conclusion had to await the twentieth century, and the discovery of the expansion of the universe by Slipher and Hubble, a development which led to the big bang theory of the cosmic origin.

A final problem about the big bounce theory is that, under a very wide range of conditions, some sort of singularity still forms (according to general relativity). True, the singularity may not block off the whole universe – that is, some of the infalling matter may miss it –

but it means that we can't avoid confronting the issue that spacetime had a boundary in the past simply by making the universe a bit lopsided. And if we have to confront a singularity anyway, we might as well make it an all-encompassing big bang singularity rather than a skulking big bounce singularity.

The foregoing account was pretty much the received wisdom when I was a student in the 1960s. The big bang was said to be an event without a cause because the spacetime singularity that bounds it signals a breakdown of both spacetime and physical theory, rendering meaningless any notions of cause and effect which might be applied to it. This seemed to place the origin of the universe forever outside science. But cosmologists were only just getting into their stride, and some adventurous theorists were determined to find a thoroughly scientific account of the cosmic birth. The way forward came from a rather unexpected direction.

How far back can we push our theories?

To try to reconstruct the story of the early universe, we have to apply our best understanding of physics to the extreme conditions just after the big bang. High-energy particle physics provides some experimental data which can be used as a guide. But as we consider earlier and earlier moments, we have to rely on increasingly speculative theories. Inflation, for example, makes use of grand unified theories (GUTs) of particle physics that so far have no direct experimental confirmation.

Even well-established theories of physics may not be applicable all the way back to time zero. Scientists use the word 'extrapolate' when they take an idea, theory or law of physics and apply it on quite a different scale of size or energy. The question here is how far our physical theories can be extrapolated back towards the moment of cosmic origin before the conditions become too extreme for us to have any confidence that they still apply unmodified. The astonishing thing about physical science is just how extendible some theories turn out to be. For example, Maxwell's theory of electromagnetism gives an excellent account of the electromagnetic properties within atoms, but also of galactic magnetic fields, some 10^{32} times larger in radius. It accounts both for the effects of minute magnetic fields on cosmic rays in intergalactic space, and the behaviour of collapsed stars called magnetars which support magnetic fields 10^{20} times stronger.

How does Einstein's general theory of relativity fare in this regard? It's true that general relativity comes out with flying colours whether applied on the scale of the solar system or the scale of the universe. But how do we know whether it will still apply when the universe is shrunk to the size of a tennis ball, or an atom? Surely it is a huge leap of faith to apply the *same theory* all the way down to zero?

Physicists have a rule of thumb about matters of scaling. If the theory has no inbuilt unit of length – nothing to fix the scale against which physical processes may be compared – then there is no way to predict when, if at all, the theory will break down. Maxwell's theory of electromagnetism has no such fundamental unit of length, even when combined with quantum mechanics. But gravitation is a different story. The general theory of relativity does not have an inbuilt unit of length, so it can scale up to the largest cosmological size or down to the smallest intervals of space and time, without giving us a clue as to when, or even whether, it might fail. But when gravitation is combined with quantum mechanics, a completely new situation arises. Max Planck, who originated the quantum theory in 1900, spotted that the new fundamental constant of physics – what we now call Planck's constant, h, the number that sets the scale of quantum phenomena – can be combined with the speed of light c and Newton's gravitational constant G (see Box 6) to make a quantity with a unit of length.[15] It is known as the Planck length in Max's honour, and its value is about 10^{-33} cm, or 10^{20} times smaller than an atomic nucleus. The existence of this fundamental length scale suggests that if gravitation is given a quantum mechanical treatment, then something important will happen when the system concerned shrinks to the Planck size. In particular, we expect that Einstein's general theory of relativity, which makes no reference to quantum phenomena, cannot be extrapolated unmodified to this situation: drastic departures from the predictions of general relativity can be expected at and below the Planck length.[16]

Another measure of when quantum effects will impinge on gravity is to construct a natural unit of time, which is obtained by dividing the Planck length by the speed of light. This yields the so-called Planck time, about 10^{-43} s. On general grounds, one might expect Einstein's theory to fail on such a small timescale, to be replaced by a quantum theory of gravity. In the context of the big bang origin of the universe and the possible presence of an initial singularity, the Planck time serves as a warning: we should not trust general relativity when extrapolated to within one Planck time of the origin. This won't affect inflation, at least in the version I have so far described, which occurs much later (at about 10^{-34} s, or a billion Planck times). But the quantum effects of gravity should dramatically alter all the stuff about singularities and the ultimate origin of the universe. It was the realization of this that led to the creation of a new field of study: quantum cosmology.

Quantum cosmology

Combining the two extreme ends of physics – cosmology, the study of the universe, and quantum mechanics, the study of atomic and subatomic systems – is ambitious by any standard. But that hasn't stopped some distinguished physicists from working on the topic. First up was John Wheeler in the 1960s, who argued that quantum uncertainty would fuzz out the singularity, replacing the infinite curvature of spacetime with something gentler and more complex. One way to see this is to compare the singularity to the point of an infinitely sharp needle. Quantum mechanics applied to a real needle would make the position of the tip slightly uncertain, in effect blunting it. Will quantum mechanics 'blunt' the spacetime singularity that may have bounded the past of our universe? Well, needles are one thing, cosmic singularities another. For a start, when dealing with quantum gravity, quantum mechanics has to be applied to *spacetime*, not to matter, raising deep technical and conceptual problems.

Even if these difficulties can be overcome, there remains the question of which quantum state the universe is in. To see the problem, think of a very simple system which needs quantum mechanics to explain it – the hydrogen atom. Quantum mechanics correctly describes how the electron orbiting the proton possesses only certain discrete energies. These energy levels start at the so-called *ground state* – the lowest energy configuration – and continue in a series of ever more energetic *excited states*. So to determine the behaviour of a given hydrogen atom, it is necessary to specify exactly which state it is in initially. For example, if the atom is in the ground state, then it would simply stay there, but if it is in one of the higher-level excited states, the electron will jump down to the ground state, emitting one or more photons. The choices for the quantum state of the hydrogen atom are unlimited (the electron may, for example, be in one of an infinite number of superpositions of energy levels – see Box 4). Likewise, the universe as a whole may be in any of a limitless number of different quantum states, and different results flow from different choices – which isn't a lot of help. So which quantum state is the universe actually in? In the early 1980s, Stephen Hawking of Cambridge University, in collaboration with James Hartle of the University of California, Santa Barbara, suggested that there might be a special 'natural' quantum state for the universe – a sort of ground state – which they expressed in terms of a particular mathematical construction.[17]

How the universe might have originated from literally nothing

Although a proper understanding of the Hartle–Hawking state requires advanced mathematics, a rough idea can be given via a picture. A key feature of the chosen state is the way in which quantum uncertainty 'fuzzes out' space and time. Applied to a particle such as an electron, quantum uncertainty implies that its position and motion are somewhat ill defined (see Box 4). Applied to spacetime, quantum uncertainty predicts that space and time themselves are somewhat ill defined: points in space get smeared, moments of time get smeared. But more than that, quantum fuzziness entails smearing the *separate identities* of space and time. Let me explain what I mean by this. In daily life, space is space and time is time – there is no ambiguity – even though space and time are *nearly* the same thing, being part and parcel of a unified spacetime. But in the quantum realm their distinct identities become blurred. Sometimes intervals of time behave like

6. *What is G?*

As I explained in Chapter 1, Newton correctly guessed that the strength of the gravitational force of attraction between two bodies diminishes with the inverse of the square of the distance that separates them. Figure 1 (p. 10) shows a graph of the force as a function of the separation. But that is not quite the whole story. Notice that there are no specific units on the graph. Newton's law tells us how the force varies *relatively* with distance (four times as great at half the distance, etc.), but it says nothing about the *absolute* strength of the force. Newton correctly deduced that the strength depends on how much matter the two bodies contain – their masses. But this is still not enough. Ask what is the gravitational force of attraction between two one-kilogram masses held one metre apart, and you can't get the answer from Newton's theory alone. The only way to find out is to measure it and see what nature has decided. Once that has been done, and assuming (as Newton did) that the law of gravitation is *universal*, then the scale is fixed and the absolute force is determined for all masses and distances everywhere in the universe. (For the curious, the answer is $6.673 \times 10^{-11} \mathrm{m}^3 \mathrm{kg}^{-1} \mathrm{s}^{-2}$.)

It is actually hard to measure the gravitational force between known masses, but it has been done. An early method was to see how much a mountain pulled a dangling weight away from the vertical, and then determine the strength of the force, knowing (roughly) the mass of the mountain. These days you can get a decent answer in the laboratory using two metal balls. The gravitational pull between them is tiny, but it can be measured with sensitive equipment. Anyway, the point is that there exists a fundamental constant of nature – Newton's gravitational constant, denoted by G – that multiplies the force law and pins down the actual value of the force. If G were twice as big, all the gravitational forces in the universe would be twice as strong (all else being equal). Let me stress again that G cannot be *derived* from Newton's theory, it has to be measured experimentally. Most physicists, following Newton, believe that G is a universal constant – that it is the same everywhere and for all time, although just why it has *that* value they haven't a clue. All the forces of nature have the same feature. Whatever the mathematical form of the force laws, there are undetermined overall constants (often called 'parameters', because their values could be different) that fix the absolute strengths of the forces and must be measured experimentally for their values to be known.

intervals of space, and vice versa: space becomes time-like and time becomes space-like. The resulting identity crises – quantum fluctuations in spaceness and timeness – are exceedingly small: in fact, they are more or less restricted to Planck lengths and times. But in the context of the origin of the universe they could be immensely significant.

To understand how the Hartle–Hawking quantum state incorporates the maybe-space-maybe-time fuzziness, look at Figure 15. This is a very schematic representation of the expanding universe. Time is represented vertically and space horizontally. I have suppressed two space dimensions, leaving only one, depicted as a circle (i.e. closed space). We can see at a glance that the universe is expanding, because the radius of the circle is bigger at later times. Conversely, in the past it is smaller, until at the beginning of time, denoted by $t = 0$, it shrinks to nothing; this is the big bang singularity. In this picture, spacetime looks like an upside-down cone, but apart from the sharp apex at the base, not too much importance should be placed on the shape. The effect of quantum uncertainty is to change the structure of the cone near the apex so that it appears more like the form shown in Figure 16. The infinitely sharp point, representing a spacetime singularity, gets replaced by a rounded bowl. The radius of this bowl is about a Planck length, extraordinarily tiny by human standards, but not actually zero – which is the crucial thing. The singularity has therefore been removed in this description.

Translated into spacetime language, and using our familiar play-the-movie-backwards description, this diagram describes a universe which contracts inexorably towards a zero radius, which it is destined

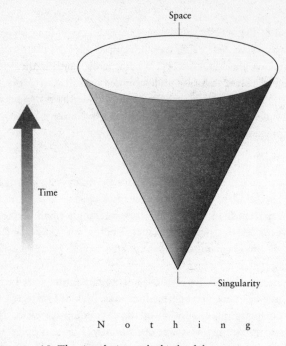

Space

Time

Singularity

N o t h i n g

15. *The singularity at the birth of the universe*

In the standard big bang model, based on Einstein's general theory of relativity combined with the assumption of perfect uniformity, the universe originates in a singular state of infinite density and infinite spacetime curvature, shown here schematically by the apex of the inverted cone. For ease of visualization I have shown space as one-dimensional, and closed into a circle (representing a three-dimensional hypersphere). The region below the cone, marked 'nothing', which seems to lie 'before' the big bang, does not exist as a physical region in this model. Both time and space start at the singularity.

to reach at a certain predicted time, but just before this singular terminating event happens (about one Planck time before it), time itself starts to get fuzzy, seized by an identity crisis, and begins to adopt more and more space-like qualities. Time doesn't change abruptly into space – the proposed Hartle–Hawking mathematical construction sees to that – but fades away in a continuous manner. At 'the base of

Space

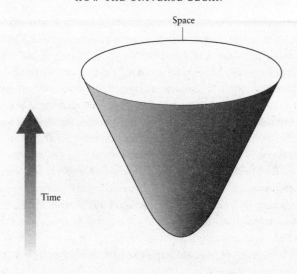

Time

N o t h i n g

16. *A quantum origin of the universe*

In this highly schematic picture, based on a proposal by Hartle and Hawking, the universe is bounded in the past, but there is no singular origin at which time suddenly 'switches on'. Rather, time becomes continuously space-like near the beginning as a result of quantum mechanical effects.

the bowl' time has become purely space-like. Adopting forwards-in-time terminology, this says that, at the start, there were actually four dimensions of space, one of which transformed itself into time. This transformation wasn't a sudden 'switching on' of time, à la Augustine, although in human terms it was rapid enough, occupying only about a Planck time (or rather it would have done, had time properly existed). But crucially, it was not instantaneous. A singular origin of the universe, the event without a cause that seemed to place the cosmic origin outside science, is replaced in this theory by a smooth origin, complying with the laws of physics everywhere.[18]

How seriously can we take the Hartle–Hawking account of the cosmic origin? Not very seriously, in my opinion. Its value lies mainly in showing us what a physical theory of the birth of the universe from absolutely nothing could be like. Whether or not the theory as formulated is correct, it demonstrates how one may deftly pass between the horns of a seemingly impossible dilemma. Before the work of Hartle and Hawking, it was assumed either that the universe must always have existed in one form or another, or that there was a first moment in time, a singular instant when time 'switched on', for no reason. The Hartle–Hawking universe, however, lets one have one's cosmological cake and eat it, because it possesses two seemingly contradictory properties. On the one hand it is finite in the past – time does not stretch back for all eternity – but on the other hand there is no well-defined instant of creation either. The base of the bowl in Figure 16 represents in some sense the past limit of time, but it is not a first moment as such.

Let me repeat that the Hartle–Hawking proposal does *not* describe a universe that has always existed. There is still a big bang, and no universe whatever existed, say, a microsecond before it. But attempts to pin down the first moment are doomed to failure, lost in the general uncertainty of quantum mechanics. Asking what came before the big bang – what lies below 'the bowl' – is futile. In Hawking's words, it is like asking what lies north of the North Pole. The answer is nothing, not because there is some mysterious Land of Nothing there, but because the epoch 'before the big bang', like the region 'north of the North Pole', simply does not exist.

Welcome though this progress was in abolishing a magical cosmic origin without falling foul of the problems of an eternal universe, the Hartle–Hawking theory, as well as various other attempts which have been made to describe the universe quantum mechanically, run up against another deep problem of principle, which is this. Either the cosmic origin is a natural event, or it is a supernatural event. (By the latter, I mean that it has no complete explanation from within science alone.) But with what justification can we declare it to be a natural event if it has happened only once? A natural event is one that can happen in conformity with the laws of nature with a probability greater than zero. Which is a careful way of saying that if a universe can pop into existence from nothing thanks to the laws of physics, it can do so again ... and again. As the Canadian philosopher John Leslie has pointed out, it would be exceedingly odd if the physical process behind the creation event bore the label THIS MECHANISM OPERATED ONLY ONCE.[19] In other words, whatever physical theory one may invoke to describe the origin of the universe, that same theory describes the origin of *many* universes – indeed, of a limitless number of universes. As it happens, just such a theory stems from the inflationary universe scenario.

Eternal inflation

One of the weak points of Guth's original inflationary universe was the need to assume that the inflaton field obligingly started out in an unstable excited state – the state that would drive exponential expansion. Calculations hinted that if the big bang that preceded inflation by a split second was hot enough, the inflaton field might automatically cool into the required state. But many theorists didn't believe this, or thought it looked pretty contrived. Two Russian cosmologists, Andrei Linde, now at Stanford University, and Alex Vilenkin, now at Tufts University, found a better way of getting inflation started which avoids appealing to special initial conditions. The basic idea is that once inflation gets going, it is generally hard to stop it happening everywhere. The reason for this has to do with the same quantum fluctuations I discussed on p. 71 in connection with the large-scale structure of the universe. The inflaton field that drives inflation is subject to Heisenberg's uncertainty principle, so its strength will fluctuate randomly and spontaneously from place to place and from time to time. The field, being in an unstable excited state, mostly wants to decay and bring inflation to a close – this was Guth's original suggestion. But in the odd spots where the field suffers a *strengthening* fluctuation greater than the rate of decay, inflation will actually pick up (the rate of inflation increases with the strength of the inflaton field). Although these against-the-trend regions are sparsely located (most fluctuations are too small to override the decay), they generate a lot of space. Remember, the hallmark of inflation is that space doubles in size in successive fixed time intervals. In terms of physical volume, then, the rare inflating regions completely dominate. So taking our by now familiar god's-eye view, the universe would consist mostly of inflating space created by rare regions with strengthening fluctuations, interspersed with regions that have ceased inflating and turned into conventional expanding universes (expanding, that is, at a decelerating rate).

Because quantum fluctuations are unstoppable, there will always be regions of space somewhere that continue to inflate, and these regions will represent the greatest physical volume of space. The system can therefore go on generating pocket universes ad infinitum. Each pocket universe will inherit the smoothness of the inflating region from which it derives, overwritten with some small quantum fluctuations that will create large-scale structure. (The fluctuations within a pocket universe are generally much smaller than the fluctuations in between one pocket universe and another.)[20] One question that is still unresolved is whether the inflating system needs to have a beginning. As in Guth's original theory, the ultimate origin could be a big bang bounded by a singularity. But it is also possible to imagine that there is no beginning: that inflating space generating pocket universes has always existed.[21] The latter proposal is preferred by Linde, who refers to his model as chaotic eternal inflation, chaotic because of the random fluctuations and eternal because inflation has no beginning and no end. 'The whole process', he writes, 'can be considered as an infinite chain reaction of creation and self-reproduction which has no end and which may have no beginning.'[22]

The multiverse

Eternal inflation dramatically changes the nature of cosmology. Although it explains the same things as do other versions of inflationary theory, it establishes a completely different conceptual basis. What we have all along been calling 'the universe' is, in this theory, just an infinitesimal part of a single 'bubble', or pocket universe, set amid an infinite assemblage of universes – a *multiverse* – itself embedded in inflating space that exists without end. Leonard Susskind, a theoretical physicist at Stanford University, refers to the multiverse poetically as 'the bubble bath universe'.[23] Eternal inflation therefore offers an inexhaustible universe-generating mechanism, of which our universe – our bubble – is but one product. Each pocket universe will be born in a burst of heat liberated in that bubble when inflation ceases, will go on to enjoy a life cycle of evolution, and will perhaps eventually suffer a death, but the bubble bath system as a whole is immortal.

Where are the other universes? The short answer is, a *very* long way away. It is a prediction of the inflation theory that the size of a typical bubble is fantastically bigger than that of the observed universe. By fantastically, I mean 'exponentially' bigger. Our observed universe is likely to be deeply embedded in a region some $10^{10,000,000,000}$ km across! Compare this with the size of the observable universe, a mere 10^{23} km.[24] And even if by some magic we could be transported to the edge of our bubble, we wouldn't encounter the universe next door. Instead, there would be a region where space is still inflating, doubling in size every 10^{-34} s or faster. So even though pocket universes like ours are expanding, they won't intersect because they are being moved apart by inflation in the gaps between them much faster than their boundaries are growing. It is thus physically impossible, even for light, to cross the widening gulf between them.

Eternal inflation represents a drastic shift not just in the science of cosmology, but in its philosophical foundations too. Suddenly, the universe has got very, very much bigger. A mere five hundred years ago, many people envisaged a cosmos, centred on Earth, which measured only thousands of kilometres across. The birth of astronomy revealed that the stars are many light years from Earth, and in the twentieth century it became apparent that other galaxies were located millions and even billions of light years away. Now this enormous extension of scale has taken another leap.

As with many scientific advances, the concept of a multiverse is not a new idea. In the seventeenth century the polymath philosopher, mathematician and physicist Gottfried Leibniz suggested that our world is but one member (the best, in fact) of an ensemble of worlds, by which he meant not merely an ensemble of planets, but of entire universes, including their own space and time, each with its own distinctive properties and arrangements of matter. In the eighteenth century the philosopher David Hume played with the idea that our universe might be the product of a long process of trial and error by an incompetent creator:

If we survey a ship, what an exalted idea must we form of the ingenuity of the carpenter, who framed so complicated, useful, and beautiful a machine? And what surprise must we feel, when we find him a stupid mechanic, who imitated others, and copied an art, which, through a long succession of ages, after multiplied trials, mistakes, corrections, deliberations, and controversies, had been gradually improving? Many worlds might have been botched and bungled, throughout an eternity, ere this system was struck out: much labour lost: many fruitless trials made: and a slow, but continued improvement carried on during infinite ages in the art of world-making.[25]

The vast increase in cosmic dimensions represented by the multiverse is only one aspect of the philosophical realignment. Since Copernicus, scientists have assumed that there is nothing special or privileged about our location in the universe. As I discussed in Chapter 2, this is often referred to as the principle of mediocrity: Earth is a typical planet orbiting a typical star in a typical galaxy. Applied to the distribution of matter in the universe, the mediocrity assumption is called the cosmological principle, which means that, in the absence of evidence to the contrary, we should assume that the universe is the same (on the large scale) everywhere. The cosmological principle is supported by the fact that the universe is uniform out as far as our instruments can measure. But eternal inflation flies in the face of the principle of mediocrity, by presenting our universe as embedded within a bubble surrounded by something very different (an inflating region). True, there may be countless other pocket universes beyond, but ours wouldn't be a humdrum bubble – far from it. As we shall see in the coming chapters, there is every reason to suppose that our pocket universe is very special indeed.

Key points

- The universe as we know it began 13.7 billion years ago with a hot big bang. The universe is still expanding today, albeit at a much diminished rate, and is bathed in heat radiation – the cosmic microwave background (CMB), the afterglow of the big bang. The CMB provides a snapshot of the universe at 380,000 years after the big bang.

- On a very large scale the universe is uniform, but at the level of clusters of galaxies and below, visible matter is clumped together. This structure is reflected in the CMB, which is smooth overall but has measurable 'ripples'.

- The basic structure of the universe is nicely explained by the theory called inflation, which postulates that in the first second of its existence the universe leapt in size by an enormous factor, caused by an intense pulse of antigravity.

- When inflation ceased, space was essentially empty. The energy of expansion was converted to heat, which brought about the creation of matter.

- Quantum fluctuations during inflation imprinted on the universe its large-scale structure.

- The big bang may or may not have been the ultimate origin of the universe. If it was, then time and space did not exist before the big bang. Cosmologists have attempted to explain scientifically the origin of the universe from nothing (no time, no space, no matter) by appealing to quantum mechanics. The resulting subject of quantum cosmology is exciting, but not rigorous.

- If the big bang was not the ultimate origin of the universe, the question arises of what came before it. In a currently popular theory known as eternal inflation, our universe is just one 'bubble' of expanding space among many, and big bangs occur throughout time in the wider 'superstructure'. Taking a god's-eye view, most of space is inflating at a fantastic rate, and the 'bubbles', or pocket universes, emerge spontaneously from this as a result of quantum processes.

- Eternal inflation is one mechanism for generating a multiplicity, or ensemble, of universes, known collectively as a multiverse. Individual universes within the multiverse could be very different from one another. Only a small fraction might be fit for life.

4

What the Universe Is Made of and How It All Holds Together

The first (credible) theory of everything

Imagine playing the role of an intelligent designer, planning a universe fit for life. The present universe works well enough, but how much could you change without spoiling things? It's possible that you could do away with some sorts of galaxy or eliminate giant black holes. Some small stars and large planets might be superfluous. At the atomic level, you could probably get rid of a few elements, but most are needed somewhere in the life story.[1] At a more fundamental level, you would be wise to leave things completely alone. Getting rid of electrons would be a disaster, as chemistry would then be impossible. Abolishing neutrons would rule out any element other than hydrogen. The inventory of fundamental particles is not a good place to tinker. Even meddling with the properties of these particles would be risky.

Given the necessity of keeping things pretty much as they are, the question arises of why the universe consists of the things that it does. Why are there electrons, protons, neutrons and all the other atomic components? Why do these entities have the properties that they do? Why do the particles all have particular masses and electric charges, and not others? Fifty years ago, nobody thought very much about such questions. Today, however, there is a feeling among physicists that we ought to be able to answer them: that the list of fundamental particles and the properties assigned to them are not arbitrary, but should be explicable in terms of a deeper theory, a theory that unifies all the disparate components. Attempts to do this are sometimes called, with considerable exaggeration, theories of everything.

Complete explanations of the physical world were offered by the fifth-century BCE Greek philosophers Leucippus and Democritus. They lived well before anything we would call science was practised, but the early philosophers were observant and skilled in the art of reasoning. Like us, they asked the big questions, such as how the universe is constructed, where it came from and what it is made of. They reached new heights of logic, mathematics and metaphysics, and came to believe that the universe could be understood by the careful and systematic application of reasoned argument.

A troublesome problem which many Greek philosophers mulled over concerned the nature of change. How did an acorn turn into a tree? How did water turn into steam? In general, how did something become something else? The difficulty, as the ancient Greeks saw it, is that physical objects possess identities (that is how we can name them). So if something is A, how can it *become* B without somehow *being* B to start with? How can a thing become what it is not? Some philosophers concluded from this perplexing paradox that all change is illusory, but others went to the opposite extreme, claiming that nothing retains a fixed identity, that everything is in a state of flux.

Leucippus and Democritus identified a clever way out of this philosophical quagmire. Suppose, they argued, that the universe consists of nothing but microscopic, indestructible particles moving in a void. The particles don't change: they are primitive, indecomposable entities, and retain a fixed identity for ever. So too with the void: it is immutable. But the *motion* of the particles within the void will bring about the appearance of change. In this scheme, all matter consists of different arrangements of particles, and all change is but the rearrangement of particles. The name philosophers gave their fundamental particles was *atomos* (constructed from *a* – 'not' and *tomos* – 'a cut'), from which we derive the scientific word 'atom'.

The atomic theory of matter was intended to give a complete and unified account of the physical universe. The indestructible nature of the atoms was crucial, for the theory depended on them having fixed identities. If the atoms could be pulled apart, then they could change, and the philosophers would be back where they started, trying to account for how one thing could turn into another. The atomists' idea was that atoms come in a variety of shapes and sizes, but within a given class they are identical. So the theory could be completed merely by assembling an inventory of the different types of atom, determining their respective shapes and sizes, and specifying how they stick together. Then, in principle, every object in the world, and every physical process, could be accounted for in terms of its atomic constituents.

Atoms today

Over the centuries, the atomic theory waxed and waned in favour. Its great advantage was its stark simplicity and sweeping explanatory power. Its main drawback was that the existence of atoms had to be accepted on faith. As they were postulated to be far too small to be seen, direct observational evidence was completely lacking. And this situation persisted right into the modern era. Indeed, distinguished scientists and philosophers were still contesting the atomic theory of matter in the first decade of the twentieth century.

Today, the existence of atoms is not in doubt. You can even see pictures of them in textbooks. In many respects, the atoms we know today resemble the particles that Leucippus and Democritus had in mind. They do indeed come in different varieties (a hundred or so) but are otherwise identical (more or less). They have different sizes and shapes, although they are all roughly spherical. They may stick to others to make molecules or crystals, and changes such as occur during chemical reactions can be traced to rearrangements of the atoms. Where the atoms we know today differ from their Greek antecedents is in their internal structure. The philosophers' atoms were supposed to be indestructible, but what we now call atoms are composite bodies with moving internal parts. Some physical processes, such as the slow alteration in the composition of the sun as hydrogen turns to helium, result from the rearrangement of the internal parts of atoms rather than of the whole atoms themselves. This complication took the shine off the atomic theory as a candidate for a complete theory of nature, once the existence of atoms had become firmly established.

For a while it was possible to hope that the essential idea of Leucippus and Democritus could be salvaged by identifying the ultimate building blocks of matter with the constituents of atoms rather than with the atoms themselves. At first, this didn't seem too much of an elaboration. An atom contains a compact nucleus consisting of a ball of protons and neutrons which possess most of the mass. The nucleus is surrounded by an extended swarm of lighter electrons. Atoms are held in this configuration by electrical forces: protons are positively charged and electrons are negatively charged, producing a force of attraction that binds the electrons to the nucleus. So could it be that the *subatomic* particles – electrons, protons and neutrons – are the indestructible objects from which all matter is built?

A host of subatomic particles

Unfortunately this hope was soon dashed. By the late 1930s several additional particles had turned up, not confined to atoms. There is a particle identical in all respects to the electron, except that it carries a positive rather than a negative electric charge. It was called the *positron*.[2] Then there is another particle which closely resembles the electron but is 207 times heavier. Physicists named it the *muon*. Moreover, muons come in both positively and negatively charged varieties, mirroring the electron and positron. And there is a ghostly particle known as the *neutrino*, with no electric charge and extraordinary penetrating power, which travels very close to the speed of light. It betrayed its existence originally merely by the energy it carried away in radioactive decay events. Neutrinos were not detected with certainty until the 1950s. By that time the situation had become even more complicated because all sorts of other subatomic entities were turning up, many of them in cosmic rays (high-speed particles that strike the Earth from space; the word 'ray' is an anachronism, but still used). For example, there is the *pion*, which also comes in both positively and negatively charged varieties, 273 times heavier than the electron, and another pion, this time electrically neutral, just a tad lighter. And there is a whole family of particles which are heavier than protons and neutrons.

So why are these other particles so elusive? The positron is a case in point. Being a sort of mirror image of the electron, a positron will disappear if it encounters an electron. The electron and positron annihilate each other and vanish, releasing their mass-energy in the form of gamma-ray photons. In isolation a positron will be perfectly stable, but because Earth is made of electron-infested matter, terrestrial positrons don't last long. Muons went undetected for a different reason. Most of them decay after a few microseconds, each negatively charged muon turning into an electron, and each positively charged muon turning into a positron,[3] so they don't exist for long enough to produce noticeable effects in the everyday world. Neutrinos weren't spotted sooner because they interact so feebly with ordinary matter that they rarely leave any trace of their existence. For example, the sun is a prolific source of neutrinos, but most of them pass straight through normal matter. Every second, your body is penetrated by billions of neutrinos with virtually no consequences. Indeed, most solar neutrinos arriving here pass through the entire Earth with scarcely a shudder, and fly off into space on the far side. These weird, elusive particles are by no means rare: in fact, they are probably the most common particles in the universe, outnumbering electrons and protons by a billion to one. But they are so inconspicuous that they might never have been discovered at all without special equipment.

By the 1960s the number of subatomic particles that had been identified was so great that physicists ran out of names and started designating them by letters and numbers instead. The subatomic world began to resemble a menagerie of weird and wonderful animals that existed for no obvious reason. Most of the particles decay in the tiniest fraction of a second, so there was no way that they could be the building blocks of familiar matter.[4] Yet they are undoubtedly a form of matter, and physicists became eager to bring some sort of order to the proliferating list of these entities.

Some patterns emerge

In spite of the bewildering number and variety of subatomic particles, some patterns became apparent early on. For example, all particles have the same electric charge (positive or negative) or none at all; the neutron and neutrino are electrically neutral, hence their names. Every species of particle has an associated antiparticle, with the same mass but with all other properties, such as electric charge, reversed. The positron was the first antiparticle to be discovered – it is an anti-electron, but is still called the positron for historical reasons. Then along came antineutrinos, and eventually antiprotons, antineutrons, and all the rest. So there is a deep symmetry here, a matter–antimatter symmetry. Another systematic property of subatomic particles is *spin*. The electron, for example, spins on its axis rather like a mini-planet, but always with the same rotation rate. For historical reasons that rate is given the unit $1/2$. Protons, neutrons, muons and neutrinos also have spin $1/2$. Other particles are known with spins of 1, $3/2$ or 2. Spins are always multiples of $1/2$, so another rule is apparent. A further important rule is that all particles of matter belong to one of two distinct categories: nuclear particles and products of their interactions (such as protons, neutrons and pions), which tend to be more massive and are collectively known as *hadrons*, and the rest (electrons, neutrinos, muons, . . .), which are lighter and called *leptons*. The words derive from the Greek for 'heavy' and 'light' respectively.

Although some of the new particles were found in cosmic rays, increasingly they came to be made in laboratories, using particle accelerators, which slam high-energy protons or electrons either into fixed targets, or into their respective antiparticles coming in the opposite direction. Mostly the accelerators are ring-shaped vacuum tubes, like the one at Brookhaven I mentioned in the last chapter (although a famous accelerator at Stanford University in California is straight). The largest accelerator today is located at the CERN Laboratory near Geneva, and measures 27 km in circumference. It was originally designed to accelerate electrons and positrons in counter-rotating beams to 99.999 per cent of the speed of light and then direct them into head-on collisions. It is now being redesigned for use with protons and antiprotons, and rechristened the Large Hadron Collider, or LHC. When it becomes operational in 2007, it will create collisions at energies corresponding to the state of the universe at a hundred-trillionth of a second after the big bang, when the temperature was close to a billion billion degrees. The purpose of these high-energy collisions is not primarily to study cosmology, but to expose the deeper structure of matter. It is a technique that dates from 1932, when John Cockcroft and Ernest Walton first used a high-voltage electrical device at Cambridge University to split an atomic nucleus.

Quarks are building blocks of matter

The ever-lengthening list of newly discovered particles consisted mostly of hadrons (the heavy, nuclear-type particles). Physicists began to think that maybe they weren't fundamental at all, but were combinations of smaller particles. Murray Gell-Mann and George Zweig suggested that protons and neutrons contained three of these smaller particles apiece, which Gell-Mann called *quarks*, after a line in a James Joyce novel. To make the scheme work, quarks must have electric charges of $1/3$ and $2/3$ of the basic unit. These two different quark varieties, or *flavours*, as physicists prefer to call them, were given the arbitrary names of *up* (the $2/3$ one) and *down* (the $1/3$ one). Two ups and a down combine to make a proton, two downs and an up to make a neutron. Naturally there is an antiquark for each quark flavour, and these can combine with quarks in unstable pairs: a pion, for example, is an up stuck to an anti-down (or vice versa, depending on whether it is a positively or a negatively charged pion). To recap, there are now *three* levels of atomic structure: atoms are made of nuclei and electrons, nuclei are made of protons and neutrons, and protons and neutrons are made of quarks. Leptons (electrons, neutrinos, . . .) are left out of this scheme: they are treated as fundamental, like the quarks.[5]

The quark scheme tidied things up a lot. It implied that commonplace matter is constructed from just four basic entities: up and down quarks, electrons and neutrinos. (The reader may think it odd that I call neutrinos commonplace because they are unfamiliar in daily life, but that is just because we don't directly sense them; as I have explained, they are more common than all the other particles put together, and you can't get more commonplace than that.) This gang of four really consists of two pairs, because the properties of the quarks differ markedly from those of the electron and neutrino.

On the face of it, the universe would tick over pretty much as it does if this quartet were its only ingredients. But for some reason, nature has not only duplicated but triplicated the above scheme. It's as if at the moment of creation someone said, 'If one family of particles is good, two is better – and three is better still!' And three is what happened. The next family has two more flavours of quark – dubbed *strange* and *charm* – and two more leptons – the muon and a second variety of neutrino. These particles are heavier than those of the first family[6]; they are unstable, and decay into particles belonging to the first family.[7] Finally there is Family 3: *top* and *bottom* quarks, a very heavy lookalike of the electron that goes by the name of *tau*, plus yet another flavour of neutrino. These particles are also unstable. To distinguish between the three flavours of neutrino, they are labelled according to the lepton with which they are paired: electron-neutrino, muon-neutrino and tau-neutrino. This scheme is summarized in Table 1. As far as we know, that's it. These three families of four seem to do the trick. The quarks can combine by mixing and matching between the three families, which makes for a lot of potential combinations and accounts for each of the large number of short-lived hadrons. And although this list of particles might still seem depressingly tedious and unmemorable, it does possess a pleasing internal consistency and order. Anyway, that's the way nature is, so we have to accept it. No doubt Leucippus and Democritus would have been satisfied.

It would be wrong for me to give the impression that particle physics is just a labelling exercise, merely sorting the subatomic menagerie into different species. The family relationships that connect the particles through the different combinations of quarks and antiquarks can all be cast in the language of mathematics (using something called group theory). Additional hidden symmetries underlie the combinations, enabling formulas to be written down which associate the properties of different particles, similar to how symmetries in geometry connect the sides of a square or the vertices of an equilateral triangle. In particle physics, though, these symmetries are not geometrical but of a more abstract nature. Nevertheless, their existence shows that the micro-world is not just a ragbag of random objects, but a harmonious realm in which the components possess deep, albeit abstract, inter-relationships. This provides a possible answer to the puzzle of what

Particles

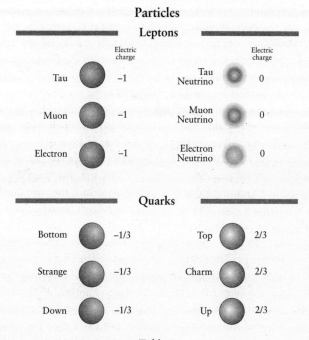

Table 1

The twelve particles that make up all known matter. They all have spin $^1/_2$. Each has a corresponding antiparticle. The quarks are not found in isolation, but exist either in combinations of three or in unstable quark–antiquark pairs. Electric charge is given here in units of the charge on the proton. The particles in Families 2 and 3 decay into Family 1 particles.

the two additional families of quarks and leptons are 'for'. It might seem that they could be thrown out without affecting the universe much. However, because the particles can be linked through various abstract symmetries, they can form part of a larger tribal grouping, which cannot be pulled to bits in a piecemeal manner. The hope is that all the particles, including the heavier unstable quarks and leptons, will eventually find a natural place in some deeper scheme (for example string theory, which I shall describe shortly).

When Gell-Mann and Zweig first proposed their quark theory, it was a shot in the dark. But experimental evidence that quarks are real accumulated in the 1970s and 1980s, and their existence is no longer in doubt. Curiously, however, nobody has ever detected an isolated quark. One of the peculiar properties of the force that binds quarks together is that it grows larger with separation. So knocking a quark out of a proton, say, is impossible. Quarks exist, but it seems that they are permanently confined to larger particles.

Successful though the quark and lepton scheme may be, it is only half the story. When the Greek atomists came up with their idea of fundamental particles, an essential part of the theory was the ability of the particles to combine in various ways. Therefore, some sort of forces must act between them. Without something to bind particles together, they would all just go their own way, and matter as we know it could not exist. The study of the inter-particle forces provides the other half of the story.

Four fundamental forces explain everything

In spite of the rich diversity of physical systems that adorns the universe, from atoms to galaxies, only four basic forces appear to be needed to explain their properties. But on the face of it, there seem to be many more forces in nature, sculpting and transforming matter on all scales. The most obvious force is gravitation – the attraction that keeps our feet on the ground. Another recognizable force is electro-static attraction – the thing that makes rubbed balloons stick to the ceiling or hair stand on end when combed. Magnets also display attractive properties, or repulsive if turned the other way round. The pressure of gas inside a balloon or steam hissing from a boiling kettle provide other familiar examples. Careful investigation, however, shows that all these forces can be traced back to just four. Atoms, for example, stick together to make molecules, or bounce apart, as a result of the electric charges they contain (it is these same charges that produce electrical attraction of the balloon-sticking variety). Everyday forces like the wind in the sails of a boat or the push of a paddle-wheel derive ultimately from this atomic electricity, averaged over vast numbers of atoms. In fact, gravitation, electricity and magnetism account for almost all phenomena in the everyday world.

In the early twentieth century two new forces were discovered. We don't notice them in daily life because they are entirely confined to the nuclei of atoms, and to close encounters between subatomic particles. One of these forces, called the strong nuclear force, or just the *strong force*, is responsible for binding quarks together to make hadrons. A remnant of this powerful inter-quark force acts between neutrons and protons and explains how atomic nuclei hold together and overcome the electrical repulsion of the protons. The force may be very strong, but it has an extremely short range, dropping sharply to zero beyond a distance of about a ten-trillionth of a centimetre. The other nuclear force is known simply as the *weak force*. It's responsible for a form of radioactivity – the phenomenon that occurs when some nuclear particles decay into others. For example, isolated neutrons are unstable, and will decay after some minutes into a proton, an electron and an antineutrino. It is the action of the weak force that causes this transmutation. The same force also explains why muons (and many other particles) decay. As far as the two nuclear forces are concerned, hadrons (i.e. quarks) respond to both the strong and weak forces, but leptons feel just the weak force.

The foregoing list seems to contain five forces – gravitation, electricity, magnetism, and the weak and strong forces. However, two of them – electricity and magnetism – are really just two faces of a single *electromagnetic force*. The link between electricity and magnetism was discovered in the nineteenth century through the work of Michael Faraday, Hans Christian Oersted, James Clerk Maxwell and others. It is very easy to demonstrate. An electric current produces a magnetic field – a phenomenon exploited in doorbells, television sets, hair dryers, and a host of other everyday gadgets. Conversely, a changing magnetic field generates electrical forces, and will make a current flow if there is a circuit. Again, this is familiar in devices such as dynamos.

In the 1850s Maxwell was able to combine the equations that describe electricity with those that describe magnetism, taking into account the interweaving of the two forces. Perhaps the most important outcome of this unification was the discovery of electromagnetic waves. Because changing magnetic fields create electric fields and changing electric fields create magnetic fields, the possibility arises of self-sustaining oscillations of electric and magnetic fields, each feeding off the variations of the other. Maxwell used his electromagnetic field equations to predict that such oscillating fields would propagate through the vacuum of empty space in the form of waves. Furthermore, the equations even supplied the speed of these waves. Significantly, when Maxwell put the numbers in, the answer turned out to be the speed of light. The conclusion was obvious: light itself must be a form of electromagnetic wave. Physicists went on to explore other electromagnetic waves – basically the same phenomenon as light, but with different wavelengths and frequencies. These range from radio waves and microwaves, through infrared radiation, visible light and ultraviolet, to X-rays and gamma rays (see Figure 7, p. 33).

Gravitation and electromagnetism can be described in terms of fields. For example, the moon orbits the Earth because the Earth's gravitational field interacts with the moon and produces an attractive force. Similarly, we say that a magnet creates a magnetic field that can reach out and interact with another magnet, producing a force. This sort of description works well on a large scale, but on the scale of atoms it fails, because there we must take into account quantum effects.

How forces work at the quantum level

Quantum mechanics requires a completely different way of thinking about how forces operate. Here is a thumbnail sketch. Suppose that two electrons are approaching each other on a collision course. Because like electric charges repel, the electrons should deflect away from each other; physicists call this process *scattering*. Quantum mechanics provides a distinctive description of how electron scattering happens, cast in the language of photons. A photon is a quantum of the electromagnetic field; it is the particle-like face of electromagnetic waves (see Box 4, p. 72). What happens when two electrons scatter is that one electron emits a photon, and the other then absorbs it. Each electron experiences a little recoil[8] as a result, and these kicks knock them onto diverging paths. The process is depicted schematically in

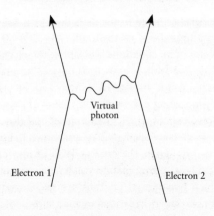

17. How quantum particles interact

This so-called Feynman diagram depicts two electrons which approach and then diverge as a result of electromagnetic forces acting between them. To describe this process using quantum mechanics, it is necessary to think of the electrons as exchanging a 'virtual' photon. When all such exchanges are considered (i.e. all possible photons), and amalgamated in the calculation to compute the net effect, spectacular agreement with experiment is obtained.

Figure 17. Photons that remain confined to a private exchange between consenting electrons (or other particles) are called *virtual*, to distinguish them from the 'real' ones that come out of light bulbs and create the sensation of light in our brains.

Pictures like Figure 17 are called Feynman diagrams after their originator, Richard Feynman. They are helpful in guiding intuition, but they should not be taken too literally. For instance, the diagram makes the scattering process look like an abrupt change in each electron's motion, but this is deceptive. Normally it isn't possible to say which electron emits the virtual photon and which absorbs it, or when. By the rules of quantum mechanics, one must take into account *all* possibilities in calculating the net effect of one electron scattering another. Every emission and absorption process contributes to the result; there isn't a single diagram corresponding to 'reality'. Attempts to pin down where and when a photon might be emitted are frustrated by the restrictions of Heisenberg's uncertainty principle, but when all processes have been included in the calculation the net result can be extracted from an amalgam of them all.[9] Even this is only part of the story, however. The equations describing the electron scattering process (which derive from a branch of physics known as *quantum electrodynamics*) cannot be solved exactly, so an approximation scheme is used, consisting of an unending series of ever more complicated calculations, each of which gives a closer and closer approximation to the answer.[10] The Feynman diagram shown in Figure 17 is just the first term in this series, the crudest level of approximation. To calculate the next best approximation you have to consider a Feynman diagram in which two virtual photons are exchanged. Still higher-order corrections to the calculations include three, four, five photon exchanges, necessitating harder and harder calculations. Again, the rules of quantum mechanics require that all possibilities – all possible numbers of photons – contribute to the net scattering effect, although the contributions (thankfully!) diminish sharply with each extra photon, so in practice theoretical physicists rarely need to go beyond second-order approximations to get accurate explanations or predictions for most problems of interest.

I have been describing electron–electron scattering, but the same theory (using the concept of virtual photon exchange) may be used to describe a host of other phenomena, such as the emission, absorption and scattering of light by charged particles and atoms, the fine details of atomic energy levels, the annihilation of electrons and positrons, and the magnetic properties of leptons. Anyway, when these calculations are performed properly, the agreement one finds between theory and experiment is absolutely stunning. Some of the predictions have been tested in experiments to an extraordinary degree of accuracy, in some cases better than one part in a trillion. Such extremely close agreement gives physicists great confidence that the 'photon exchange' explanation of quantum electrodynamics is correct.

In this quantum description, the existence of the electromagnetic interaction between two charged particles is attributed to the exchange of another particle – the photon. The electromagnetic field of a charged particle may be envisaged as a cloud of virtual photons surrounding the particle. A similar scheme applies to the other forces. Gravitation operates by the exchange of *gravitons*. Nobody has ever detected a graviton directly, because gravity is such a weak force, but its properties can be deduced from what we know about gravitational fields. The weak nuclear force needs *two* exchange particles, called W and Z (actually, W is electrically charged, and comes in both positive and negative forms, denoted by W$^+$ and W$^-$). The strong nuclear force is more complicated still, and requires no fewer than eight exchange particles to glue the neutrons and protons together; these quanta are collectively known as *gluons*. So to complete the inventory of how the universe is put together, we may add to the list of six quarks and six leptons (plus their antiparticles) a total of twelve exchange particles responsible for the forces that act on them. This is summarized in Table 2.

The weak and electromagnetic forces are united

If nature used a legion of forces, physicists would be inclined to just list them and be done with it. But four is a curious number: one might expect either one force or very many. It prompts the question of whether the four forces, very different though they may be in their properties, are actually one and the same force manifested in four different guises. As I have explained, electricity and magnetism have long been known to be linked. Could it be that the remaining forces are in some way connected, deep down? It is a beguiling idea with a long history. In the 1850s Michael Faraday unsuccessfully tried to find a connection between electricity and gravitation by dropping heavy weights and measuring their electrical effects.

In the event, further unification of the forces came from a quite different quarter. In the 1960s physicists became intrigued by certain similarities between electromagnetism and the weak nuclear force. On the face of it, these forces are unpromising bedfellows. The weak force is extremely short in range, being confined to subnuclear dimensions, whereas electromagnetic fields can extend across an entire galaxy. The electromagnetic force is also very much stronger than the weak force. But viewing the action of the forces quantum mechanically – as an exchange of virtual particles – brings them closer. Physicists had long known that the range of a force is directly related to the mass of the

BOSONS

Unified electroweak (spin = 1)		
Name	Mass GeV/c^2	Electric charge
γ photon	0	0
W⁻	80.4	−1
W⁺	80.4	+1
Z⁰	91.187	0

Strong (colour) (spin = 1)		
Name	Mass GeV/c^2	Electric charge
g gluon	0	0

Gravitation (spin = 2)		
Name	Mass GeV/c^2	Electric charge
graviton	0	0

Table 2

The four forces of nature are conveyed by exchange particles with spins 1 and 2. The eight different gluons are here shown as one entry because they perform a similar function. The graviton has never been detected in an experiment; its existence and properties are inferred from mathematical theory. The units used for the masses of W and Z are the conventional technical ones (known as GeV/c^2); for comparison, the proton mass in the same units is 0.9383. Electric charge is given in units of the charge on the proton.

virtual particle that is exchanged: the bigger the mass, the shorter the range. The photon is massless,[11] so electromagnetism has unlimited range. The fact that the weak force has such a short range implies that it exchanges virtual particles of very high mass. If somehow the mass of the weak force's exchange particles could vanish, then the two forces are actually quite similar in their properties – similar enough, at least, to write down a common mathematical scheme to amalgamate them.

Such a scheme was worked out by Sheldon Glashow, Steven

Weinberg and Abdus Salam. They proposed that the weak force is conveyed by the three high-mass particles, W^+, W^- and Z, just discussed. The basic idea is that, 'underneath', the weak and electromagnetic forces are really an amalgam of forces on an equal footing, but this common source is concealed by the high masses of the W^+, W^- and Z. Crucially, however, the scheme works only if one assumes that W^+, W^- and Z are 'really' massless – that they have no *intrinsic* mass – but instead acquire their *effective* (high) masses by a new type of physical interaction called the Higgs mechanism. I shall have more to say about the Higgs mechanism for generating mass in Chapter 7. For now, I ask you to accept that the Higgs mechanism removes the obstacle to a unified description of the two forces, enabling a satisfactory account of both to be encompassed within a single set of equations.

But that is not all. The theory also explains the huge disparity in the strengths of the forces. Basically, the two forces are equally strong 'underneath', but to obtain the low effective strength of the weak force the theory demands that you divide the energy of the participating particles by the masses of W and Z. As these masses are very large, the effective strength of the weak force at low energies is strongly suppressed. However – and here the theory makes a clearly testable prediction – as the energy is raised, so the weak force should grow stronger. Raise the energy enough and the two forces – weak and electromagnetic – will be revealed as two faces of a blended *electroweak force*. One way to raise the energy is to collide particles together at high speed. In 1983, the CERN laboratory tested the electroweak theory by colliding protons and antiprotons at energies equivalent to dozens of proton masses, and were rewarded by discovering both the two W particles and the Z particle that the Glashow–Salam–Weinberg (GSW) theory had predicted.

The Standard Model

Meanwhile, theorists had not overlooked the strong nuclear force – the one that binds the quarks together. A scenario was worked out by analogy with quantum electrodynamics, the theory of how charged particles interact via photons. It goes like this. The quarks carry some sort of strong force 'charge' to make them interact with a 'strong' field. The strong charge is called *colour* (there is no connection with the normal meaning of the term). To get the theory to work you need three colours (in contrast to a single form of electric charge). And in place of one photon you need eight exchange particles, called gluons, to hold the quarks together. A theory along these lines, known as *quantum chromodynamics*, or QCD, was proposed in 1973 by David Gross, David Politzer and Frank Wilczek, and it beautifully accounts for the known experimental data about hadrons.

From the point of view of experiment, this is as far as we have come. The advances I have been describing – the quark theory, leptons, the electroweak unification – have been consolidated into what is now called 'the Standard Model of particle physics'. It describes twelve quarks and twelve leptons (as shown in Table 1) and incorporates the GSW theory of the electroweak force and the QCD theory of the strong force.

The Standard Model has been tested in many ways over the past thirty years, and has always passed with flying colours. Nevertheless, physicists are adamant that it is not the final say on the subject. For a start, some experiments already hint at new physics beyond the Standard Model, physics which should be clarified when the LHC accelerator starts operating. Then there are some important facts which the Standard Model doesn't say anything about, such as the need for three families of quarks and leptons, and the existence of dark matter and dark energy (which I shall describe in Chapter 6). Moreover, the Standard Model has the air of unfinished business. One ugly feature is the tentative way in which it bundles together the electroweak and strong forces, like apples and pears in the same basket, without attempting to combine them. The Standard Model looks like a halfway house to a fully unified theory in which the strong and electroweak forces would be merged into a single superforce.

Another shortcoming of the Standard Model is that the fourth force in the quartet, gravitation, is left out of the account completely. Although all particles feel the gravitational force, it is so incredibly weak that gravitational processes at the subatomic level are completely overwhelmed by the effects of the other forces. Also, gravitation is something of the odd man out because it can be described, not as a force, but as a warping of spacetime geometry. This geometrical feature makes it much harder to bring gravitation within the sort of quantum description that is well adapted to the electromagnetic, weak and strong forces. Nevertheless, it would be odd if there were two forces of nature – a unified electroweak/strong force, and gravitation. The successive unification of particles and forces hints at the existence of a more embracing unity, one that combines *everything* into a single mathematical scheme. The lure of a final theory – a theory of everything – has proved irresistible to a generation of theoretical physicists. And now, some enthusiasts claim, they could be on the verge of attaining their goal.

Key points

- Matter is made of atoms, atoms are made of electrons and nuclei, nuclei are made of protons and neutrons, and protons and neutrons are made of quarks. There are many additional particles, but most of them decay very rapidly. Most of them are combinations of quarks and antiquarks.
- Four fundamental forces – gravitation, electromagnetism, and the weak and strong nuclear forces – are enough to explain how all known matter behaves. At the quantum level, forces are described as the exchange of virtual particles.
- Particles of matter divide into quarks and leptons. Quarks feel the strong force, leptons don't.
- The four forces are probably linked. Two of them, the electromagnetic and weak forces, have been successfully combined as an 'electroweak' force.
- The Standard Model of particle physics juxtaposes the unified electroweak theory with a theory of the strong force called quantum chromodynamics (QCD). It is immensely successful in explaining what we know about particle physics. There are, however, some important facts that remain unaccounted for by the Standard Model, so physicists regard it as a first step towards a more embracing theory.

5

The Lure of Complete Unification

Grand unified theories

All science is a search for unification. Science as we know it today began when Newton, Galileo and others found links between the motion of bodies on Earth and the movement of the moon and planets. Other landmark connections were the discovery that magnetism and electricity are related to each other, and to light, and Einstein's formula $E = mc^2$, which showed that energy and mass are equivalent. Finding hidden links between seemingly disparate phenomena is what makes the scientific method so powerful and compelling. The distinctive feature of science is that it is both broad and deep: broad in the way it tackles all physical phenomena, and deep in the way it weaves them, economically, into a common explanatory scheme requiring fewer and fewer assumptions. No other system of thought can match its breadth and depth.

The thrust to go beyond what is now the Standard Model of particle physics was already apparent in the 1970s, which is when the term *grand unified theory*, or GUT, came into use. Following the success of the electroweak unification theory (i.e. the GSW theory), several GUTs were published purporting to unify the electroweak force with the strong nuclear force, using the same basic ideas as GSW, but now bringing in the eight gluons from QCD too. GUTs made a clear and arresting prediction: the differences between all three forces should diminish as the energy is raised. However, the *scale* of unification is very different indeed. Calculations indicated that the three forces converge to equal strength when the energy reaches about ten trillion times that of the electroweak unification energy (see Figure 18).

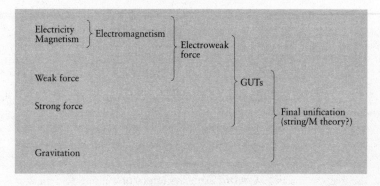

18. *Unification of the forces*

Physicists think that all the forces of nature could be linked, perhaps in a unifying scheme such as string/M theory. Historically, electricity and magnetism were the first to be unified (by Maxwell in the 1850s). Then the electromagnetic force was combined with the weak force in an electroweak theory by Glashow, Salam and Weinberg (GSW theory), and confirmed by experiment. Various grand unified theories (GUTs) have set out ways to unite the strong force with the electroweak force, but so far there is no experimental confirmation. Gravitation is brought into the unification programme as a final step, and is included in fashionable theories such as string theory and M theory.

Unfortunately, achieving such energies seems a daunting task. In fact, the GUT unification scale lies beyond even the highest-energy cosmic rays, which makes these theories very hard – but not impossible – to test experimentally. For example, some GUTs allow for a tiny degree of mixing, or spillover, between the strong and electroweak forces. One effect of this would be to permit transmutations among particles that would be forbidden under the Standard Model: on very rare occasions, for example, quarks could change into leptons or vice versa. One manifestation of such a transmutation would be protons spontaneously decaying into positrons. Experiments have been done to look for this, but without success. If such a decay does ever happen, it must have an average lifetime in excess of 10^{35} years, or it would have been spotted by now.[1] Proton decay would have profound implications for the ultimate fate of the universe, for it implies that, over immense durations, all normal matter will slowly evaporate away. Gas, dust, rocks, planets and even the burned-out remnants of stars would totally vanish in the fullness of time.

What happened to all the antimatter?

The possibility that protons might not be absolutely stable is also germane to the origin of the universe. If matter can disappear, then it can also appear (by the reverse process). This yields a clue to one of the deepest puzzles of cosmology: the origin of matter. Somehow it was made, in a flash, from the heat energy of the big bang. But cosmologists want to know *exactly* how it happened, and why that particular amount (10^{50} tonnes in the observable universe) got made. When matter is made in the lab by high-energy collisions, the same quantity of antimatter appears too.[2] If the universe contained equal amounts of matter and antimatter we'd be in for trouble. Whenever antimatter and matter mingle, they quickly annihilate in a burst of gamma rays. Even in outer space lots of mingling happens, when gas clouds collide, for example. Unless matter and antimatter are quarantined on a very large scale (much larger than the size of a galaxy), the universe would be flooded with distinctive gamma radiation. Astronomers have looked for it, without success, and so have concluded that less than one-millionth of our galaxy is in the form of antimatter. Most cosmologists assume that the entire observable universe is made overwhelmingly of matter. So that presents a puzzle: how did the big bang make 10^{50} tonnes of matter without also making 10^{50} tonnes of antimatter?

Evidently the symmetry between matter and antimatter cannot be exact. Something must have broken it, slightly favouring matter over antimatter. Grand unified theories naturally encompass the necessary symmetry-breaking. If a proton can turn into a positron, then (by going the other way) an electron–positron pair can turn into an electron–proton pair – in effect, a hydrogen atom with no anti-hydrogen atom accompanying it. But however it is done, the story of the origin of matter would go something like this. The heat radiation released after the big bang created copious quantities of both matter and antimatter, all mixed together, but containing *a slight excess* of matter. As the universe cooled, the antimatter would be totally destroyed by virtue of its being in intimate contact with matter, leaving unscathed the small residue of excess matter – about one part in a billion. The wholesale annihilation of the antimatter, and most of the matter, flooded the universe with gamma-ray photons. Where are they now? The answer is that they lost most of their energy as the universe expanded and cooled, eventually becoming microwave photons. They constitute the cosmic microwave background radiation. So this radiation is a fading remnant of the primordial extermination visited on antimatter at the dawn of time.

Viewed this way, matter is almost a cosmic afterthought. But what an important afterthought it is! Without matter there could be no life. Therefore our very existence, not to mention the existence of the visible universe, hinges on the minute degree of symmetry-breaking between matter and antimatter, which in turn depends on how quarks, leptons and the forces that act between them are amalgamated together in some as yet undetermined grand unification.

Supersymmetry

It's easy to see why physicists have been bitten by the unification bug. The plethora of subatomic particles has been successfully organized around the tidy system shown in Table 1 (see p. 107), and the four forces have been reduced to three, or maybe two, with a single super-force beckoning (see Table 2, p. 114). Convergence is the name of the game. But theorists have long set their sights on an even more ambitious unification project, a scheme that unites *particles* with *forces*. This beguiling prospect is opened up by the fact that forces can be described in terms of the exchange of particles, as depicted in Figure 17 (see p. 111). So why not cook up a theory in which *all* the particles – matter particles and exchange particles – are amalgamated in a giant superfamily?

Before this step can be taken, a very basic difference between the two classes of particle must be confronted, a difference concerning their spins. All fundamental particles of matter, the quarks and the leptons, possess spin $1/2$. By contrast, the known exchange particles all have spin 1 (the graviton, if it exists, would have spin 2). This may seem a mere technicality, but it's not. The spin of a particle deeply affects its properties, especially in the way the particles behave collectively. Particles with spin $1/2$ obey a rule known as *Pauli's exclusion principle*, after Wolfgang Pauli who discovered it in the 1920s. It forbids more than one particle of each type from occupying the same quantum state at the same time (e.g. you can't squash electrons too closely together), and it has far-reaching consequences: the solidity of matter, the structure of atoms, the rules of chemistry and the stability of stars, to name but a few.

The Pauli principle doesn't apply to the exchange particles that convey the forces, because these have a spin of 1 or 2; there is no limit to how many particles of this sort can crowd together.[3] In a laser beam, for example, countless photons occupy the same quantum state. Particles with integer spin (0, 1, 2, . . .) are known as *bosons* (after Satyendra Bose), while particles with half-integer spin ($1/2$, $3/2$, . . .) are called *fermions* (after Enrico Fermi). Uniting matter and forces thus entails unifying fermions and bosons, and these seem to be such different animals that the task looks hopeless from the start. But in 1973 a way round this obstacle was found. To get the gist of what it involves, you need to have some conception of the nature of a particle's intrinsic spin. A particle with spin $1/2$ behaves weirdly when its axis is rotated (it is possible to rotate the spin axes of subatomic particles by applying a magnetic field). Think of a macroscopic spinning body such as a planet. If it is rotated by 180° about a line through the equator (perpendicular to the spin axis), then the north and south poles would be transposed. Rotate it a further 180° and the starting configuration would be restored. That much is obvious enough; now comes the weird part. If you perform the analogous operation on an electron (or any spin $1/2$ fermion) you have to rotate it through 720° – i.e. *two* entire revolutions – before it returns to its original state! This is another one of those impossible-to-visualize properties of quantum physics, but there is no doubt it is correct. Somehow, fermions have a sort of double view of the world. Bosons, however, don't share it. The key to providing a unified theory of fermions and bosons rests with finding a geometrical description that combines

Supersymmetry

Spin 0	Spin 1/2	Spin 1	Spin 3/2	Spin 2
Higgs	Higgsino			
Slepton	Lepton			
Squark	Quark		Gravitino	Graviton
	Gluino	Gluon		
	Photino	Photon		
	Zino	Z		
	Wino	W		

Table 3

A pleasing mathematical scheme called supersymmetry predicts that every known type of particle should be accompanied by a partner with different units of spin, in such a way that every fermion has an associated boson and vice versa. No supersymmetric partner particle has yet been detected, indicating that nature cannot be exactly supersymmetric.

both forms of rotation in one mathematical scheme. Such a scheme was discovered by Julius Wess and Bruno Zumino, and is called *supersymmetry*.

If supersymmetry is present in nature, it makes a sweeping prediction: for every species of fermion there should be a corresponding supersymmetric partner that is a boson, and vice versa. Thus electrons should be twinned with so-called selectrons, their supersymmetric, spin 0 partners. (A spin 0 particle is a boson that has no intrinsic spin.) There should also be squarks, sneutrinos, and so on. Conversely, photons get twinned with spin ½ particles called photinos. Then there should be winos, zinos, gluinos and gravitinos (the latter being fermions with spin ³⁄₂). I have set all this out in Table 3. It is very fascinating – except that to date, no supersymmetric partner of any sort has been found. But this doesn't mean that supersymmetry is wrong. If it were an exact symmetry of nature, then selectrons would have precisely the same masses as electrons, winos the same mass as W's, and so on. But if some physical mechanism were to break supersymmetry, it could make the masses of all the supersymmetric partners enormous. It would then be no surprise that these partner particles have not yet been created in accelerator collisions (although they may be present in cosmic rays or dark matter). Many physicists are pinning their hopes on CERN's Large Hadron Collider, which they predict will make at least one supersymmetric partner particle. If so, it would confirm that matter and force are indeed merely two aspects of a single underlying supersymmetric scheme.

Catastrophe at the bottom

So far in this chapter I have used the word 'particle' in a rather cavalier way, without saying anything about the size or shape. This issue must now be confronted. How should we envisage, say, an electron? It is tempting to think of it as a little ball with electric charge smeared throughout it. If it were really like that, then it opens up awkward questions, such as what the stuff inside the electron is made of and how it is held together, especially as electric charge is repulsive and will try to tear the little ball apart. Obviously, if the electron could be dismembered, it would not be a truly elementary particle after all.

One way around this would be if the electron were perfectly rigid, like a totally solid miniature golf ball – that would make it indestructible. There is a problem with this proposal, however, having to do with the theory of relativity (which applies even at the level of subatomic particles). Imagine striking a golf ball with a club, sending it flying. Being perfectly rigid, the golf ball would have to move off without any change of shape: all parts of the ball must start moving together. But now we run into a snag. No force can travel faster than light, so the blow delivered to one side of the ball can't be felt by the other side until at least the time that light would take to traverse the ball. Consequently the struck side would have to start moving before the remote side. But then the ball would have to change shape – it would be compressed. It follows that the ball must have at least a certain amount of squashiness: *perfectly* rigid bodies are inconsistent with the theory of relativity. But if an electron can be squashed, then it can also be stretched – and, if assaulted violently enough, pulled apart. So a little golf-ball electron couldn't be a truly elementary body either.

But what if we imagine the little ball shrunk to a single point? Light would then take no time at all to traverse the (zero) distance across it. Unfortunately that solves one problem only to create another. There is electric charge distributed through the little ball. Imagine trying to shrink the ball, complete with its resident charge, to a smaller and smaller radius. To compress the charge into ever smaller volumes requires the expenditure of energy to overcome the electrical repulsion. According to the inverse square law of electric force discovered in the eighteenth century by Charles Coulomb,[4] the repulsion between the parts of the ball rises without limit as the charge is confined to ever smaller volumes. An infinite amount of energy would be needed to compress the ball to zero radius, and this energy would be stored inside the electron. Taking into account Einstein's formula $E = mc^2$, an infinite internal energy has the nonsensical implication that the electron should have an infinite mass. So we are left with a dilemma: the electron can be neither a point nor a finite ball without coming into flagrant contradiction with reality.

Now, you might think that quantum mechanics would come to the rescue here. By smearing out the spatial location of a point-like particle, it would seem to circumvent the difficulty that all portions of the electric charge are accumulated at a single place. In fact, quantum mechanics makes the problem even worse. To get some idea of why, remember how electric forces are conveyed in quantum mechanics – by the exchange of photons (see Figure 17, p. 111). The same forces will also act *between* the various bits of charge distributed through the 'little ball', implying that a swarm of virtual photons surrounds and interpenetrates the electron. A calculation shows that the swarm's energy gets bigger as the size of the electron gets smaller, because the close-in virtual photons are the most energetic. The total energy of the swarm rises to infinity as the radius of the electron is shrunk to zero. It doesn't matter that the overall spatial location of the electron might be ill defined: wherever it is, the cloud is there with it, clothing it with limitless amounts of energy, and hence mass.

What are we to make of this? By using mathematical tricks, physicists are able to dodge round the infinities and still use the theory of quantum electrodynamics to obtain sensible answers to questions about particle masses, energy levels, scattering processes, and so on. The theory remains brilliantly successful. But the fact that infinities occur is a worrying symptom that something is deeply wrong, something that needs fixing.

The same general analysis can be applied to the gravitational field. Shrinking a ball to zero radius would involve infinite gravitational energy. (I have already described, in Chapter 3, the spacetime singularity that would follow.) Quantum mechanically, the gravitational force is conveyed by gravitons, and the gravitational field surrounding a particle can be envisaged as a cloud of virtual gravitons. As in the electromagnetic case, infinities follow. But with gravitation there is double trouble. Any point-like particle (e.g. an electron) would be surrounded by a virtual graviton cloud containing infinite energy. But because energy is a source of gravitation, gravitons themselves contribute to the total gravitational field. (In effect, gravity gravitates.) So each virtual graviton in the cloud surrounding the central particle possesses its *own* cloud of yet more gravitons clustering around *it*, and so on ad infinitum: clouds around clouds around clouds . . . and each cloud contains infinite energy! This time the infinities can't be so easily dodged. A straightforward quantum description of the gravitational field produces a limitless progression of infinities, ruining any hope of making sensible predictions from the theory.[5]

Strings: a theory of everything?

The problem of infinities, and especially those that arise from applying quantum mechanics to the gravitational field, dogged the subject of particle physics for decades, but in the 1980s physicists found a way round it. The basic idea is to abandon the notion of particles altogether and replace it with flexible *strings*, moving according to the rules of quantum mechanics. In the simplest version of the new theory, the strings form closed loops, but they are so tiny that it would take a chain of a hundred billion billion of them to stretch across a single atomic nucleus. So what we previously took to be a particle – for example, an electron – is actually (according to this theory) a loop of string, only we don't see it that way because the loop is so small.

The attractive thing about string theory is that you need only one sort of string to make *all* the particles: fermions and bosons, matter particles and exchange particles – the whole lot. The string can vibrate in different patterns, each pattern corresponding to a different particle: if the string wriggles this way it's an electron, that way, it's a quark, and so on. String theory offers a natural description of all known particles. It incorporates supersymmetry, so it also describes the various supersymmetric partner particles. (For this reason it is sometimes referred to as superstring theory.) Don't ask what the strings themselves are made from: the whole idea is that they are primitive, indecomposable entities out of which everything else is built. In this respect, strings are very close to the spirit of the original atomic theory of matter, but they go one better. They also explain how the particles interact without introducing a separate concept; the forces come out of string theory too because there are also string motions describing the various exchange particles such as the photons and the gluons. Crucially, string theory has produced finite results in all calculations performed so far, including those involving gravitation, thus promising to avoid the mathematical gobbledygook of infinite quantities infesting the theory.

The strings are so small that it's hard to see how they could be directly observed. In this respect, we are in a similar position to the Greek philosophers who postulated atoms without having any hope of actually seeing them. To reveal the details of a string would demand a particle accelerator trillions of times more powerful than anything so far built. Even the highest-energy cosmic ray is still a billion times too feeble. Lack of direct confirmation that strings exist would not be too problematic if the theory made clear predictions about the relatively low-energy world we *can* access with current technology, but it doesn't (at least, not yet). So string theory deals with a rarefied domain of ultra-high energies and ultra-small distances, and doesn't so far have much to say about the real physics that takes place in real laboratories.[6]

Another issue concerns the space in which the strings move – the stringy equivalent of Democritus' void. The hope is that space and time will emerge from string theory as part of its description of reality, but this has not yet been achieved. You have to assume that space and time already exist to provide an arena for the strings to move in. But it's actually worse than that. In the simplest formulation of string theory, it is necessary to introduce extra dimensions of space. That is, our normal three dimensions of space have to be augmented by several others. Because we are not aware of the additional space dimensions, we have to invoke a mechanism to hide them – the process called *compactification*, which I described in Chapter 2. In itself, compactification of extra dimensions, although hard to visualize, is not a serious problem for the theory. A thornier issue is the fact that the shape and topology of the compactified dimensions are not unique. In fact, that is an understatement. Even a few extra dimensions can be compactified in a huge number of different shapes and topologies, and each arrangement leads to different particles and forces in the remaining (uncompactified) three-dimensional world.[7] According to string theory, our world corresponds to just *one* such compactified shape. But which one? And what about all the others? What sorts of worlds do they describe? As far as can be told, they would be very different from the world we observe. Some might have ten rather than three species of neutrino, or five sorts of photon. Others might have only four quarks, or as many as forty. There could be worlds in which electromagnetism is stronger than the strong nuclear force, or where there are eight fundamental forces instead of four – and so on. Obviously our world is just one possibility among a dizzying number of alternatives. Given that the goal of string theory is to unify nature, it would seem to be a step backwards if the theory predicts a vast number of alternative worlds.

String theorists are sharply split on how this multiplicity issue will pan out. Some pin their hopes on a better understanding of the mathematical structure of the theory, which they believe could single out a unique state, analogous to the ground state of a hydrogen atom, say – the most stable state and therefore the most likely. If they are right, then the low-energy physics described by this special state had better correspond to the world we observe – or the theory is immediately falsified. But other string theorists have abandoned hope that a single solution to the equations will emerge, and have boldly faced up to the consequences of an awesome proliferation of different possible worlds. In fact, they have turned a sin into a virtue. As I shall explain in the next chapter, they have invoked the vast multiplicity of possible worlds in an effort to explain the Goldilocks effect.

M theory

String theory ran into another problem when it was found that there isn't just one theory, but five. For a while this multiplicity of theories, in addition to the vast number of alternative compactification configurations, looked set to derail the whole enterprise. But just as people were beginning to lose heart, it was rescued, and from an unexpected direction. In the mid-1980s a small group of theorists had suggested that maybe strings moving in ten dimensions (nine of space and one of time) could more elegantly be described as sheets, or membranes, moving in eleven dimensions (a sheet, when rolled up like a drinking straw, would look like a string).[8] For years the membrane idea was sidelined, but in the mid-1990s Joe Polchinski of the Kavli Institute of Theoretical Physics in Santa Barbara found that in theories in which strings were not closed into a loop, but had open ends, the ends terminated on membranes.[9]

This was the circuit-breaker. A brilliant mathematical physicist, Ed Witten of the Institute for Advanced Study in Princeton, showed that a membrane description united the five different versions of string theory. He called this revivified project *M theory* – M for membrane, mystery or magic, according to taste. Mystery is apt, because the mathematical structure of M theory remains baffling and unexplored. The five previous string theories are like five 'corners' of M theory where calculations can be performed, but nobody has yet written down the equations that govern the full M theory, let alone solved them. In spite of this murkiness, M theory has generated tremendous enthusiasm. Discoveries have been made about the mathematical structure of the theory which are so surprising and suggestive that they are described by some as nothing short of 'miraculous' (M is for miracle too). These mathematical fragments provide a tantalizing glimpse of a still unexplored theory of extraordinary power and elegance that may yet turn out to be the key to the universe.

Although M theory undoubtedly represents progress, it uses branches of mathematics which are not only extremely abstract, but also extremely obscure. In fact, some of the mathematics had to be invented along the way. The fact that it is so hard leaves most physicists (certainly this one) far behind, and it leaves string/M theorists without much of a reality check. Where this enterprise will end is anybody's guess. Maybe string/M theorists really have stumbled upon the Holy Grail of science, in which case one day they might be able to tell the rest of us how it all works. Or maybe they are all away in Never-Never Land. Time will tell. Michio Kaku, himself a string theorist, wistfully expressed his concerns in a recent article: 'If string theory itself is wrong, then millions of hours, thousands of papers, hundreds of conferences, and scores of books (mine included) will have been in vain. What we hoped was a "theory of everything" would turn out to be a theory of nothing'.[10]

Whatever the outcome of this monumental project, it at least deserves the description 'nice try'. And it is certainly too soon to write it off as just another theorists' bandwagon, for to date it offers the best hope for producing a final unified theory. But one observation is worth a thousand ingenious theories, and while string theorists were frantically developing their abstract models, astronomers made a series of discoveries that exploded like a grenade in the entire theoretical physics paradigm, throwing string theory as well as cosmology into turmoil.

Key points

- Various grand unified theories try to combine the strong, weak and electromagnetic forces in a single scheme. Some predict that protons will decay with an extremely long average life.

- The symmetry between matter and antimatter cannot be exact, or the big bang would have coughed out equal amounts of each.
- Fermions and bosons can be united within a mathematical scheme called supersymmetry.
- Complete unification of particles and forces, including gravitation, is achieved by treating all particles as composed of tiny strings moving in ten spacetime dimensions. The six unobserved dimensions are rolled up in a complicated shape.
- String theory, and its further development as M theory, offers the most promising hope for unifying all of fundamental physics, but the theory remains incompletely understood and is hard to test experimentally.

6

Dark Forces of the Cosmos

Dark matter

For centuries, astronomers thought that they were studying 'the universe' when they turned their instruments on galaxies, stars and planets, and gas and dust. It therefore came as a rude shock when they found that most of the universe is made of Something Else – and scientists haven't the foggiest idea what it is!

The first inkling that in cosmology what you see isn't necessarily all there is came over seventy years ago with the meticulous work of Fritz Zwicky, a Swiss astrophysicist who worked at the California Institute of Technology. Zwicky was aware that the universe is expanding in an orderly way. But he realized that this sparse description must be a simplification. Galaxies are not isolated bodies: they congregate into clusters of maybe a few dozen members. Within the clusters individual galaxies roam about, so superimposed on the uniform expansion are complicated local movements. For example, our own Milky Way galaxy and the Andromeda Galaxy are heading towards each other at about 130 kilometres per second, while at the same time participating in the overall cosmological expansion discovered by Hubble. Zwicky was interested in studying these localized motions within clusters, which he could probe by carefully measuring the red shift in the light from each individual galaxy.

What he found was decidedly odd. Most galaxies seemed to be moving unexpectedly quickly. Zwicky initially assumed that the clusters are held together by the gravity of their visible material. If a galaxy moves too fast, it will escape the attraction of its neighbours and wander away from the group. A cluster of galaxies can remain bound together for billions of years, but only if it contains enough material to trap the individual members. Zwicky calculated that the combined gravity of all the visible material – stars, gas and dust – was nowhere near sufficient to confine such swiftly moving galaxies in a cluster. The conclusion was inescapable: there must be an extra contribution to the gravitational pull. Putting in the numbers, Zwicky discovered that the discrepancy was huge: the unseen gravitating material outweighed the visible stuff hundreds of times, completely dominating the masses of the clusters. This unseen and unknown material became known as *dark matter*. Although Zwicky's results were ignored for a long time, over the past few decades astronomers have amassed irrefutable evidence that the luminous parts of galaxies represent just the tip of a mysterious iceberg, and that most of the matter in the universe is in fact dark.

The existence of dark matter is also apparent in the way stars orbit within the Milky Way galaxy. The sun, for example, takes about 250 million years to complete one circuit. The telltale signs of dark matter come from studying how stars move on the periphery of the galaxy. Once again, they are found to be travelling much too fast to remain gravitationally bound to the Milky Way unless there is a lot of dark matter tugging on them. If the stars were all there is, the Milky Way would speedily unravel like an exploding flywheel. Astronomers have mapped the distribution of mass in our galaxy and others from the way stars move, and what they have come up with is that the familiar disk-shape structure – with a blob-like nucleus and spiral arms enfolding it – sits in the middle of a roughly spherical distribution of dark matter that stretches far beyond the fringes of the luminous region, forming a vast cloud, or halo, continuously extending into intergalactic space.

These mapping studies have been augmented by X-ray satellite observations and the results from the WMAP satellite. All the investigations point to the same conclusion: there is much more dark matter in the universe than visible matter. Obviously scientists are keen to know what it is, and there is certainly no lack of candidates. Astronomers divide them into two broad categories: MACHOs and WIMPs.

MACHOs

MACHO stands for 'massive compact halo object' – some concentration of mass residing in the galactic halo. Some MACHO candidates immediately come to mind. Black holes are dark, and go unseen unless they find themselves in the vicinity of stars or gas that they suck in. Dwarf stars or giant planets may be too dim to show up in telescopes, but still exist in abundance. Then there are smaller objects, such as asteroids and comets, which are undoubtedly very abundant but are extremely inconspicuous outside the solar system.

Spotting a MACHO in deep space is very difficult, for obvious reasons. Direct telescopic searches for small, dim, red stars have not turned up a huge number of them. Another method that has been tried is gravitational lensing (see p. 42). If a MACHO should by chance interpose itself directly in the line of sight to a star, it will betray its presence by amplifying the starlight. So a wandering MACHO will show up as a distinctive rise and fall in the luminosity of a distant star. A few MACHOs have been found in this way, but astronomers are now convinced that there are nowhere near enough of them to account for all the dark matter.

Cosmologists can get another lead on dark matter from an entirely different line of reasoning. As I described in Chapter 3, during the first few minutes after the big bang, nuclear reactions transformed hydrogen into helium. This came about when protons and neutrons in the primordial plasma combined to form, first, deuterium (one proton stuck to one neutron) and then, by the further fusion of deuterium nuclei, helium. But tiny quantities of deuterium never made it as far as helium, and got left over. The final abundance of primordial deuterium depends critically on the density of the universe at the time these reactions were taking place. Deuterium has a rather weakly bound nucleus, easily destroyed by collisions with protons. A high-density universe would lead to more frequent collisions between nuclei, thereby reducing the deuterium abundance in the final mix. Conversely, a low-density universe would leave a larger fraction of unincorporated deuterium detritus. So a measurement of the abundance of deuterium and other light elements[1] can be used to put a figure on the density of nuclear matter in the early universe, and hence – by a simple scaling argument – the density of ordinary matter today.

According to the best estimates of deuterium abundance, the early universe had a relatively low density of nuclear matter. In fact, *barely a few per cent* of the dark matter can be in the form of ordinary atoms or their components. So this rules out MACHOs as an explanation of dark matter if they are made of ordinary stuff – that is, electrons, protons and neutrons. Some MACHOs might conceivably be made of an unknown type of matter – non-nuclear matter – that didn't participate in the production of deuterium and helium. But if we are off into the land of hypothetical particles, then the theorists are eagerly waiting with a long list. And most of these would not be good for MACHOs, because they are WIMPs.

WIMPs

WIMP stands for 'weakly interacting massive particle'. We have already met a weakly interacting particle which is a possible candidate for dark matter – the neutrino. Neutrinos are not so much dark as in black but dark as in invisible, given that they mostly pass through ordinary matter without betraying their presence. And neutrinos are extremely abundant in the universe, greatly outnumbering nuclear particles. However, they don't weigh much, possibly only about a millionth of an electron. So even with a billion-to-one numerical advantage, the neutrinos wouldn't be enough on their own to outweigh the stars. Hence the importance of the M in WIMP: *massive* is what we need – something like a neutrino, perhaps, but with the mass of a proton or more.[2] That way they could dominate over the gravitational effects of normal matter without us noticing them. Like neutrinos, WIMPs could be streaming through our bodies in vast numbers all the time without us being aware of the fact.

There are plenty of *hypothetical* particles that fit the bill – for example, supersymmetric particles such as the photino. In fact there are too many contenders, leaving experimenters not quite knowing what to look for. The main problem in setting out to detect WIMPs is that, by definition, these particles would interact only exceedingly weakly with matter. Calculations suggest, however, that very rarely a WIMP will be stopped by the nucleus of an atom and deposit some energy. The challenge is to detect the tiny resulting bump and filter the signal out from the background noise. One method that has been tried is to use a large pure crystal of germanium as a detector, and to look for the effects of a recoiling nucleus, either from its electrical or its acoustic disturbance (in the latter case, the experimenters listen for the bump). The crystal is taken deep underground to filter out the much more strongly interacting cosmic rays that would swamp the signal from a WIMP. Astronomers who have mapped the distribution of dark matter believe that it tends to concentrate near the centres of galaxies. They envisage a thick invisible soup of WIMPs through which Earth and sun must swim on their long journey round the Milky Way. If that is so, the WIMPs won't be peppering Earth evenly from all directions, but should be streaming towards us from the constellation Virgo, which is where the solar system is currently heading at about 300 km per second.

Dark matter plays an essential role in shaping the universe, by providing most of the gravitational pull needed to grow galaxies. The universe at 380,000 years, as revealed by WMAP, was extremely smooth; the large-scale structure has emerged because the slightly over-dense regions were able to draw in surrounding material and thereby amplify their density. If it were down to normal matter alone, this process would have been too feeble to make galaxies, stars, planets, and so on, without which life would be impossible. But dark matter greatly assisted the clumping process. To find out how, cosmologists have compared the observed large-scale structure of the models with the results of elaborate computer simulations that model a variety of dark matter components.[3]

Although cosmologists still have few clues about the nature of dark matter, at least they can put an accurate figure on how much of it there is in total. WMAP findings, combined with the results from telescope surveys and other data, indicate that, of the entire mass content of the universe, ordinary matter (protons, neutrons, electrons, atoms, molecules) makes up only about 4 per cent of the total (and of that, only about half is in the form of stars and planets). Thus fully 96 per cent of the universe is composed of some mysterious dark stuff.

That in itself is remarkable enough. But a further surprise lies in store. Of this 96 per cent, less than a third can consist of the type of dark matter candidates I have already described. What makes up the rest – at least two-thirds of the mass of the universe – is something far more baffling.

Dark energy

In the mid-1990s two groups of astronomers stunned the scientific community by announcing that the expansion rate of the universe is actually speeding up, as indicated by observations of supernovas in distant galaxies. That is, the universe is now expanding faster than before, and looks set to run away with itself if the trend continues. The discovery rocked the foundations of cosmological theory, built as it was on the firm conviction that gravitation acts as a brake on the expansion, serving to slow it down from its explosive start at the big bang to the relatively modest rate observed today. Now the name of the game had changed. A mysterious antigravity force is opposing gravity and has succeeded in transforming deceleration into acceleration.

We saw in Chapter 3 how inflation – a huge acceleration in the expansion rate during the very early universe – was driven by a pulse of antigravity created by the negative pressure of a hypothetical 'inflaton field'. It now seems that the universe (at least, its observable portion) has begun inflating again, but at a relative snail's pace, fifty-odd powers of ten slower than it did during the very early universe. What is going on?

In 1917 Einstein proposed a universal repulsion, or antigravity force, as a way to concoct a static universe (Box 3, p. 66). He abandoned it when he found that the universe is in fact expanding. Well, he may have been right all along. To be sure, the universe isn't static, but it seems as if the antigravity is there anyway. If the correct explanation lies with Einstein's 1917 theory, the expansion of the universe won't look like the orthodox picture shown in Figure 12 (p. 59). Instead, it will resemble Figure 19. Here, the universe still originates with a big bang. In the early stages antigravity doesn't have

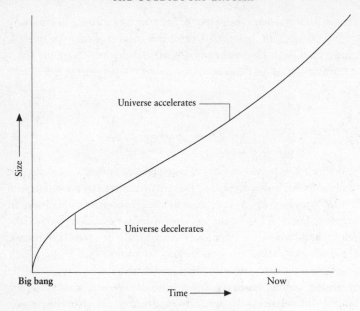

Size →

Big bang

Now

Time →

19. *The runaway universe*

Astronomers think that the expansion of the universe is accelerating, under the action of an antigravity force. The graph shows the best guess for the cosmic evolution, starting with a big bang followed by a few billion years of decelerating expansion, similar to the behaviour shown in Figure 12. But then the expansion rate begins to pick up as the universe becomes dominated by a mysterious dark energy.

much effect, because the universe is very compressed and the repulsive force postulated by Einstein is weak over small distances. But as the universe continues to expand, so the antigravity grows in strength, until a point is reached at which – over the scale of the universe as a whole – it rivals the normal attractive force of gravity. A titanic tussle then ensues, with these immense forces roughly balanced out: the universe coasts for a while with an almost constant expansion rate. Inevitably, however, the antigravity eventually wins because it continues to gain in strength the more the universe expands. Once antigravity predominates, the rate of expansion starts to pick up, getting faster and faster all the time, until eventually it approaches

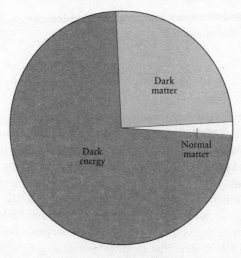

20. *A cosmic afterthought?*

Surprisingly, familiar matter such as atoms makes up only about 4 per cent of the mass of the universe – almost an afterthought. About 22 per cent is in the form of another sort of matter, as yet unidentified, while 74 per cent is in the form of dark energy that pervades all of space.

'exponential' expansion, in which the size of any given region of the universe doubles every few billion years.

It is too soon to proclaim that the force causing the universe to accelerate is one and the same as Einstein's original antigravity, although that is certainly the simplest explanation. Other suggestions have been made, for example fields in which the strength of the force varies with time. As I have explained (on pp. 66 and 68), antigravity can be considered as a consequence of the energy – and the concomitant negative pressure – of empty space itself. Alternatively, we can attribute the energy and negative pressure to an invisible field that permeates space. Either way, we don't see anything of it, so the generic term *dark energy* is used to denote all these possibilities. Astronomers are planning better measurements to find out more. Whatever it is, if you add up the dark energy responsible for making the universe accelerate, you find that it actually represents a total mass that is more than all the matter – visible and dark – put together (see Figure 20).

It seems that dark energy constitutes most of the mass of the universe, yet nobody knows what it is.

The end of the universe

The nature of dark energy certainly has momentous consequences, for nothing less than the ultimate fate of the universe hinges on it. A longstanding issue for cosmologists has been whether the universe will go on expanding for ever, or will slow down in its expansion rate to the point where it will start to contract. Einstein's general theory of relativity permits both possibilities, according to the amount of matter the universe contains. Leaving out the dark energy for the moment, it has been known since the work of Alexander Friedmann in the early 1920s that there are three distinct scenarios. First there is a low-density universe. Here, the explosive power of the big bang is sufficient for the cosmological material to overcome its own gravity, and for the universe to continue expanding. The rate of expansion slows as a result of the braking effect, but the deceleration decreases with time until eventually the universe expands at a near-constant rate. This is depicted by the curve marked a in Figure 21.

The second possibility, illustrated by curve b in Figure 21, is a high-density universe. It contains more matter, leading to a more powerful gravitational pull and a stronger braking effect. The expansion progressively slows until it halts completely, after which the universe starts to contract, falling back in on itself. The pace of contraction accelerates until it turns into a headlong collapse, popularly referred to as 'the big crunch'. The third possibility, curve c, lies on the borderline between the previous two. Here the rate of expansion progressively diminishes, but never to the point where it stops altogether. General relativity links the three cases with the geometry of space. In b, the high gravitational field of the dense matter curves space into a hypersphere. In a, space curves negatively and is open and infinite. In c, space is flat and infinite.

The three simple options a, b and c become complicated if dark energy is included. As I have already mentioned, the behaviour of the universe in the early stages is little affected by the antigravitating effect

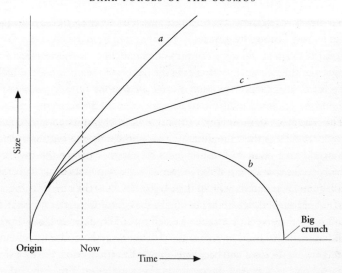

21. *The fate of the cosmos*

Three alternative models of the universe, due to Alexander Friedmann. The universe will either expand for ever at a nearly constant rate (*a*), collapse to a big crunch (*b*), or hover on the borderline between them (*c*). In *a*, space is curved negatively (see Figure 10, p. 48). In *b*, it is curved positively (see Figure 8, p. 40). In *c*, space is flat.

of dark energy, but eventually it can prove decisive. If the universe has a certain critical total mass-energy density (including the dark energy), then space is flat, and the universe behaves as shown in Figure 19. This is the model that best fits the data. The far future of the universe then looks rather bleak. Accelerating expansion turns the universe into a sort of inside-out black hole. In Chapter 2, I explained how the finite speed of light leads to the existence of a horizon in space beyond which we cannot see, however powerful our instruments. Accelerating expansion creates another type of horizon, called an *event horizon*, similar to the surface of a black hole. Imagine that a pulse of light is emitted from Earth and directed towards some distant, retreating galaxy. As the light chases the galaxy, so the galaxy recedes farther. If the expansion rate is steady, the light pulse would reach the galaxy eventually. But if the expansion rate is accelerating, the light may *never* reach the galaxy: as fast as it tries to close the gap, so the gap widens further. By symmetry, light emitted from the distant galaxy towards Earth at the same cosmic epoch may likewise never reach us, however long we wait. In that case the region of the universe in which the retreating galaxy is situated (and all regions farther away still) would be rendered invisible to us – for ever. In an accelerating universe, the galaxies move apart from one another faster and faster, and eventually disappear completely across one another's event horizons. It would take many billions of years for most of the galaxies we see today to disappear, but if the universe really does behave as depicted in Figure 19, it will happen in due course. By then the nearby galaxies (e.g. Andromeda), which are bound to the Milky Way gravitationally, will have merged into a super-galaxy full of monstrous black holes and burnt-out stars. The rest of the visible universe (still many billions of light years across) will be all but empty. Eventually even the monster black holes will evaporate away into heat radiation, which will then vanish across the horizon, along with everything else.

The big rip

Eternal expansion versus collapse to a big crunch does not exhaust the possibilities: there is a third way for the universe to end. I have assumed so far that the dark energy is constant in space and time, as in Einstein's original theory. Since the nature of dark energy is still a complete mystery, this assumption may not be correct. If dark energy is due to a new sort of matter field (which is sometimes referred to as *quintessence*), then the field would quite probably vary with space and time, and even interact with matter, leading to more possibilities. For example, there could be a sudden quantum transition to a lower value of dark energy, creating a bubble of vacuum that would expand at near light speed and engulf the observable universe. Stephen Baxter made this scary scenario the theme of his science fiction novel *Time*.[4] Alternatively, the dark energy might vary very slowly, causing it to fall in strength gradually over billions of years. Eventually it may go negative (when it would start to behave like gravity rather than antigravity). If that happened, the accelerating expansion would slow and then turn over to become accelerating collapse – a big crunch would ensue.

An altogether more dramatic fate lies in store if the dark energy *increases* in value as the universe expands. The *rate* of acceleration would then also increase with time, leading to super-exponential expansion. The event horizon would shrink, reducing the volume of space within the visible universe. The antigravity force, which is currently minuscule over solar-system or even galactic dimensions, would start to become significant on smaller and smaller scales of size. The time would eventually come when our galaxy is ripped apart. Still the antigravity would rise in strength, tearing into star clusters and eventually individual stars, overwhelming the gravity that holds them together. In the final stages even Earth would be rent asunder. Moments later, the very atoms of the universe would explode. The final act of this drama would come when the expansion rate becomes infinite (see Figure 22). Such a state of affairs represents a spacetime singularity – an end of space and time – like the big crunch, but in reverse, because it is catastrophic expansion rather than collapse. John Barrow and I discovered this unsavoury fate for the universe in the 1980s, but we didn't take it too seriously because it was based on a rather artificial mathematical model.[5] However, a few years ago the same essential idea was rediscovered by Robert Caldwell of Dartmouth College, New Hampshire, and given the catchy name 'the big rip'.[6] It is still a rather unlikely way for the universe to end, but the theory is perhaps not as speculative as when Barrow and I first toyed with it.

Can life survive for ever?

Whichever model turns out to be closest to the truth, we have a vast amount of time before Armageddon. Nothing that astronomers and cosmologists have discovered leads them to believe that the demise of the universe by any of these scenarios will begin for billions upon billions of years. However, long before anything unpleasant happens on a grand scale, our sun will have passed its use-by date. Probably in a little over a billion years from now it will have grown

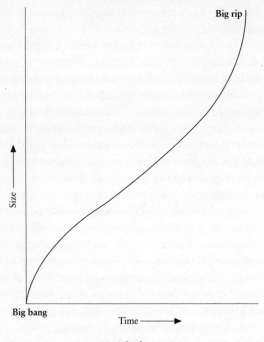

22. *The big rip*

Another way in which the universe might end is by expanding faster and faster until the expansion rate hits infinity, as shown in this graph. The curve does not continue beyond the point of infinite expansion rate, because that is a spacetime singularity.

uncomfortably hot, and Earth will face a habitability crisis. But, if one takes an unrestricted view of the future of life, there is ample time for our descendants, or any artificial organisms we may create – or for that matter any beings in other star systems faced with a similar fate – to decamp to new star systems, and continue doing so until the supply of raw material for new stars is exhausted. Even when all the stars have burned out, an even greater stock of energy will be available in gravitational fields, especially those of black holes, to enable life of some sort to continue.

But can life go on literally *for ever* (assuming that the universe won't collapse to a big crunch or explode in a big rip)? Can our descendants somehow eke out an existence for all eternity? Following a trailblazing paper by Freeman Dyson in the 1970s,[7] the popular view has been that life can always retain a toehold somewhere in an ever-expanding universe. Recently, however, that conclusion has been revised in the light of the discovery of dark energy. If, as that discovery suggests, the universe is going to end up as empty space within an event horizon, then it seems unlikely that any form of life, or even any systematic information processing, could continue, because the end state is one of thermodynamic equilibrium, similar to the original heat death of the universe discussed in the nineteenth century (see p. 83). So one way or another, it looks as if life cannot survive for ever within this universe.[8]

In this chapter I have discussed various theories about what dark energy might be, but I haven't said anything about what determines its overall magnitude. Einstein simply picked a value drawn from astronomical observations that would enable his equations to describe a static universe: general relativity provides no clue as to what the number should be. So for decades, the antigravity force strength was an arbitrary parameter which most cosmologists preferred to set to zero. But if dark energy derives from a physical process, such as quantum vacuum energy or quintessence, then theorists can have a crack at calculating the amount of dark energy in a typical volume of space. As we shall see in the following chapter, when this is done, the result is so unnerving that it signals nothing less than a crisis at the heart of theoretical physics, and the demand for the biggest fix of them all.

Key points

- Most of the universe is made of something that is still not identified. Ordinary matter makes up just a few per cent.
- Dark matter is probably made of heavy, weakly interacting particles coughed out of the big bang in profusion.
- It looks as though most of the mass of the universe is in the form of 'dark energy' – big shock! Dark energy (not to be confused with dark matter) is gravitationally repulsive, making the universe

expand faster and faster. Nobody knows what it is. It could just be the energy of empty space (Einstein suggested this in 1917), or it could be a mysterious new field.

- If dark energy stays constant, the universe will eventually become dark and empty, expanding at an exponential rate. If it diminishes and eventually goes negative, the universe will collapse to a big crunch. If it grows, the universe will end in a big rip.

- It will be hard for life to hang in there for ever and ever. Amen.

7

A Universe Fit for Life

The role of observers

A landmark in the history of science – perhaps the very birth of science itself – was the posthumous publication of Nicolaus Copernicus' book *De revolutionibus orbium coelestium* in 1543. In this tract the Polish astronomer set out his model of the solar system, with the sun at the centre and Earth and the other planets orbiting around it: 'Finally we shall place the Sun himself at the centre of the Universe. All this is suggested by the systematic procession of events and the harmony of the whole Universe, if only we face the facts, as they say, "with both eyes open".'[1]

The new view of the cosmos that Copernicus advocated, with Earth revolving around the sun, so shook the Western world that it resulted in the word 'revolution' being adopted in a broad political and social context too. In the sixteenth century, Europe was in thrall to the Roman Church, which clung to the ancient Ptolemaic view that the Earth lay at the centre of the universe and – the natural corollary – that humankind represented the pinnacle of God's creation. By demoting our planet from the pivotal cosmic position, Copernicus initiated a trend – a principle of mediocrity – that continued for four centuries. When Galileo turned the newly invented telescope on the sky in 1609 he saw that the Milky Way consisted of a vast aggregation of faint stars. Astronomers gradually came to realize that the sun is just one unexceptional star among many. Today we know that the Milky Way galaxy contains over a hundred billion stars, many of them similar to the sun. In the twentieth century, large telescopes were able to resolve stars in the Andromeda Galaxy and beyond, revealing that even the Milky Way doesn't occupy a special location in the universe. Systematic surveys of the galaxies established the cosmological principle, which proceeds from the basis that the universe is uniform on a large scale.

Any attempt to explain the universe can claim success only if it accounts for this 'mediocratic progression'. As we have seen, the inflationary universe scenario offers a natural explanation for the large-scale cosmic uniformity. The theory of how galaxies form from primordial irregularities goes some way towards explaining the ubiquity of systems like the Milky Way, while the theory of how stars and planets form from clouds of gas and dust likewise accounts for our apparently run-of-the-mill solar system. But uniformity and mediocrity are by no means the only features of the universe that must be explained. There is one aspect that often gets left off the list of observed properties, and this is the fact that there are observers to observe them.

The role of 'the observer' in science is a peculiar one, and makes many scientists a little queasy. After all, it is the task of science to replace a subjective view of nature with an objective one. A scientific claim is taken seriously only if it can be tested by others in a disinterested (not uninterested!) way. If I assert that the Earth goes round the sun because it is towed by a giant spacecraft that I alone can see, nobody would believe me – and rightly so. Einstein's theory of relativity elevates objectivity to a central principle. The very word 'relativity' implies that the view of the world is always the view of a given observer, and the theory provides the necessary transformation rules to reconcile one observer's experiences with another's. In this way the 'objective essence' of nature can be extracted from the specific experiences of individual observers. Einstein made it a founding tenet of the theory of relativity that the laws of physics must be the *same* for all observers, no matter how they move or where they are located. No observer is privileged. For this reason the laws of physics can make no reference to the speed of an object through space, for example. If it did, it would single out a certain class of observers – stationary ones – who experienced the world differently.

Given this history, it is no surprise that an explanation of the existence of observers has not been a priority requirement among cosmologists for a successful theory of the universe. But more and more of them are now convinced that this is a serious omission. Take our location in the universe – just how typical is it? Well, there is a very obvious sense in which it is extremely *atypical*. Most of the universe is near-empty space, but humans live on the surface of a planet. There is a good reason for this. Life is most unlikely to emerge in outer space, and even if it did, organisms with big brains wouldn't be very successful there. Human observers find themselves living on a planet because they could hardly have evolved anywhere else.

This rather trivial example is a pointer to more weighty considerations. Observers – at least in our experience so far – are living organisms, and life is a delicate and complex phenomenon which has many special requirements. It will emerge in the universe only if the circumstances are right. If those circumstances were not universal, then our view of the cosmos would not be typical: it would reflect our situation in a special, life-encouraging cosmic locale.

Nobody can be at all surprised by this simple tautology. It merely says that observers will find themselves located only where life can exist. It could hardly be otherwise. Nevertheless, in spite of the incontestable nature of the statement, it should not be dismissed as mere wordplay. For a start, the conditions necessary for life could be very restrictive. To be sure, Earth may be a typical planet near a typical star in a typical galaxy as far as its geological and astronomical circumstances are concerned, but it may be highly atypical – even unique – from the point of view of its biological circumstances. If there were just one place in the universe where life was able to form, Earth would be it, because that is where we find ourselves located. This point, while glaringly obvious, is in direct contradiction with the principle of mediocrity, and has become known as the *anthropic principle*.[2] The term is an unfortunate misnomer, because 'anthropic' derives from the same Greek root as 'man', and nobody is suggesting that the principle has anything to do with humans per se (though human beings are undeniably one example of life). The British astrophysicist Brandon Carter, who first used the word in this context,[3] once remarked that had he known the trouble it would cause, he would have suggested something else – the 'biophilic' principle, perhaps. But we seem to be stuck with 'anthropic', so I shall continue to use that term.

Are we alone in the universe?

What can be said about the possibilities for life beyond Earth? In spite of the huge growth in the subject of astrobiology, there is still no direct evidence for any extraterrestrial life. In the event that life is discovered elsewhere in the solar system, for example on Mars, then the most likely explanation will be that it did not originate there, but was transplanted from Earth in rocks ejected from our planet by comet and asteroid impacts. We know that Earth and Mars trade rocks, and it seems very probable that microbes have hitched a ride many times throughout the solar system's long 4.5 billion year history.[4] So finding life on Mars would not in itself prove that it has formed from scratch more than once. To draw that stronger conclusion, it would be necessary to demonstrate that Mars life and Earth life were sufficiently different to have had independent origins.

The question of whether or not we are alone in the universe is one of the great unsolved puzzles of science. The answer hinges on whether the origin of life was a stupendous chemical fluke that may have happened only once in the observable universe, or was the expected outcome of intrinsically bio-friendly laws that facilitate the emergence of life wherever Earth-like conditions prevail. For both points of view there are distinguished scientists who will argue the case, but in the absence of any hard evidence it is impossible to say either way. The stuff of life is certainly widespread in the universe. Life (at least as we know it) is based on carbon, one of the most common elements. It also uses hydrogen – the most common element of all – as well as nitrogen, oxygen, sulphur and phosphorus, all of which are reasonably abundant. Some of the building blocks of life, organic molecules such as formaldehyde and urea, are found in interstellar clouds. Water – another indispensable requirement for terrestrial biology – is extremely common in the solar system and in other star systems and gas clouds. Accumulating evidence therefore suggests that life-encouraging substances – organics and water – are found throughout the universe. Nevertheless, it is a long road from simple building blocks to even the most primitive living organism. Whereas the presence of organics and water are certainly necessary prerequisites for life, they are far from sufficient. It isn't possible to conclude, merely from the commonplace nature of life's building blocks, that life *itself* is commonplace. But as far as we can tell, life could have formed almost anywhere in the observed universe. If life turns out to be restricted to Earth, that is probably just a historical accident rather than an indicator that there is something exceptional about the solar system's biological suitability.

To permit life in at least one place in the universe, three very basic requirements must be satisfied:[5]

1. The laws of physics should permit stable complex structures to form.
2. The universe should possess the sort of substances, such as carbon, that biology uses.
3. An appropriate setting must exist in which the vital components come together in the appropriate way.

Even these three requirements impose very stringent restrictions on physics and cosmology, so stringent that they strike some scientists as nothing short of a fix – a 'put-up job', to repeat Fred Hoyle's colourful term. In this chapter I shall give some examples[6] of these striking 'cosmic coincidences', beginning with the one that so astonished Hoyle.

The origin of the chemical elements

Chemists have identified over a hundred chemical elements, which are neatly marshalled into rows and columns in the famous periodic table originated by Dmitri Mendeleev in the nineteenth century. Some elements, such as carbon and iron, are common, while others, such as gold and lanthanum, are rare. In the universe as a whole, 99 per cent of ordinary (as opposed to dark) matter is in the form of hydrogen and helium, the helium having been made during the first few minutes after the big bang. One of the challenges to astrophysical theory is to account for the relative abundances of the remaining elements. For a long time their origin was a mystery, but by the 1940s it had become clear that stars are deeply implicated. A star is a gigantic nuclear reactor, hot enough to synthesize heavier elements from lighter ones.

The first stars began to form when hydrogen and helium gas left over from the big bang aggregated into dense clouds. This took several hundred million years, because the cosmic background radiation was initially too fierce to allow the gases to settle. Eventually, under the pull of gravity, the condensing clouds fragmented into protostars – balls of gas that grew hotter as they shrank. Once the centre of a ball of gas reaches a temperature of a few million degrees, nuclear reactions are triggered. A true star is born when this heat creates enough internal pressure to arrest the contraction. Thus stabilized, a star will burn happily for millions or even billions of years, depending on its mass. Low-mass stars are relatively cool, so they consume their nuclear fuel slowly and live for a long time. Massive stars gobble up their fuel quickly and soon burn out.

Stars derive most of their energy from the conversion of hydrogen into helium by nuclear fusion. In the big bang, this transmutation was fast and easy because the universe was born with a plentiful supply of free neutrons. (Helium, remember, has a nucleus containing two protons and two neutrons. A hydrogen nucleus consists of a single proton.) But neutrons are unstable in isolation, so there weren't any on the loose by the time stars began to form. Another, much slower route to helium had to be found, using just protons. Protons repel one another because they all carry a positive electric charge, but at high temperatures they can move so fast that they can be brought into close proximity. If the speed is great enough, as it is in the hot core of a star, two protons can approach each other to within the range of the strong nuclear force, opening the way to nuclear transformations. Remember that the strong force falls sharply to zero beyond about a ten-trillionth of a centimetre, which is roughly the size of an atomic nucleus, so only by getting very close will protons come under its influence. When they do, the nuclear force is strong enough to over-whelm the longer-ranged electrical repulsion. The details don't need to be spelt out here; suffice it to say that in the process of forming one helium nucleus from four protons, two of the protons must turn into neutrons.[7] Although it is the strong nuclear force that is responsible for the prodigious energy released by fusion reactions, the crucial transmutation from protons into neutrons is controlled by the *weak* nuclear force. Because the latter force *is* so weak, it greatly slows down the entire helium production process – compared, that is, with the rate of the primordial synthesis of helium just after the big bang. This is fortunate, because it permits the great majority of stars to burn steadily for an immense duration; in the case of the sun, for long enough to permit the formation of life and to nurture the evolution of complex organisms.

When a star's supply of hydrogen runs low, it faces an energy crisis. Stars of small and medium mass can't generate any more heat from nuclear processes. They shrink to form white dwarfs, which glow only with remnant heat. Higher-mass stars, however, can call upon further nuclear fusion reactions to keep them shining, because they can reach higher internal temperatures (up to hundreds of millions of degrees). So what is the next step after hydrogen fusion? The direct way ahead would be to add another proton to helium to make lithium. However, this reaction won't work because a lithium nucleus with three protons and two neutrons is unstable; lithium normally has either three or four neutrons. What about the fusion of two helium nuclei to make the isotope beryllium-8, a nucleus with four protons and four neutrons? No good either, because that nucleus is also highly unstable, disintegrating almost as soon as it forms. The stable isotope of beryllium found in nature has five neutrons, not four. So the star is confronted with a serious nuclear bottleneck.

How the universe made carbon, the life-giving element

After beryllium, carbon is the next-heaviest element. It has six protons and six neutrons. Could it be that stars have found a way to vault over lithium and beryllium and go straight from helium to carbon? This would require *three* helium nuclei to come together at the same moment. The proton and neutron arithmetic ($3 \times 2 \times 2 = 6 + 6$)

works out correctly, and the end product would be stable carbon nuclei. Because more protons are involved in a triple nuclear encounter than in the original hydrogen fusion, the electrical repulsion is correspondingly greater, so the temperature must be higher to overcome it and allow the nuclei to get close enough for the short-range strong nuclear force to act. That isn't a problem: by further contracting, a star's core can raise the temperature to a high enough level. There is, however, a fundamental difficulty with the reaction itself. The likelihood of three helium nuclei coming together at the same place and the same time is tiny. To be sure, they don't have to arrive at *exactly* the same moment; two helium nuclei could first form a very unstable nucleus of beryllium, and before it fell apart a third helium nucleus might slam into it. But at first sight the numbers look very unfavourable, with a typical beryllium nucleus disintegrating too quickly to give a third helium nucleus a decent chance to hit it. On the face of it, then, that route to carbon seems to be blocked too.

That was the situation as it presented itself to astrophysicists in the early 1950s. Fred Hoyle, then a relatively unknown English astronomer, took an interest in this enigma. He reasoned that carbon-based sentient beings in general, and Fred Hoyle in particular, would not exist if the synthesis of the elements had become stuck at helium. Well, it's obvious enough that *something* must have happened to make carbon, presumably something inside stars. And if a general consideration of nuclear physics fails to account for carbon, then perhaps something unusual is responsible.

Which gets us to the heart of the matter. In science, one tries to avoid appealing to flukes. Occam's razor entreats us to try the simple and obvious explanations first. But sometimes simple and obvious explanations just won't work, and we are forced to invoke something out of the ordinary. As Sherlock Holmes remarked, when you have eliminated the impossible, then what remains, however improbable, must be the truth. On the whole, sticking to the simple and the obvious is the best strategy, but there is one topic where even extraordinary flukes can enter into a legitimate scientific explanation – and that topic is life.

To get what I am driving at, consider this: *not one of your ancestors died childless*. In the history of humanity, infant mortality has mostly been very high. There have been huge numbers of children who have died before they reached puberty. Now imagine your pre-human ancestors, stretching back hundreds of millions of years.[8] A long time ago your ancestors were fish. Think how fish spawn countless eggs, and imagine the tiny, tiny fraction that survive and mature. Nevertheless, not one of your ancestors – not a single one – was a failed fish. What are the odds against this sequence of lucky accidents extending unbroken over billions of years, generation after generation? No human lottery would dare to offer such adverse odds. But here you are – a winner in the great Darwinian game of chance! Does this mean that there is something miraculous in the history of your ancestry? Not at all. If your very existence depends on a concatenation of freak events, then such events may become part of a perfectly valid scientific explanation. Scientists call this an observer selection effect. By looking at the world through the eyes of an observer, the world you see will have to encompass whatever it takes for you, the observer, to be seeing it. This 'anthropic' reasoning was applied by Hoyle to the problem of carbon synthesis in stars by appealing, not to a freak sequence of events as with our ancestors, but to an unexpected and fluky property of atomic nuclei.

Here's how it happens. The rate at which nuclear reactions proceed depends on the energy of the participating particles. Mostly the variation in the rate is a gentle rise or fall in efficiency, but occasionally there is a sharp spike in the rate. Physicists call this abrupt amplification a resonance. The name stems from the way that quantum mechanics enters the picture. Quantum theory ascribes a wave aspect to particles (see p. 72), including atomic nuclei, and waves famously display resonances. For example, some opera singers can deliver a high-frequency note that resonates with a wine glass, enough to shatter it. A more mundane example of resonance comes from tuning a radio to receive the signal from a particular station. When the frequency of the circuits in the radio matches the frequency of the radio waves from the station, the waves resonate with the circuit and the signal is greatly amplified. Quantum waves can also resonate, thereby boosting the rate of an atomic or nuclear process.

Hoyle felt that resonance held the key to an explanation for carbon production. The mass of a normal carbon nucleus is rather less than the masses of the three helium nuclei that it was supposed might collide to form it, because of the mass-energy released when the carbon is made. But nuclei can exist in excited states too, so Hoyle deduced that a carbon nucleus must have an excited state a little bit above the combined mass-energies of three helium nuclei. The helium–beryllium system could then resonate at this mass-energy if the small deficit were made up by the kinetic energy of the particles rushing about inside the hot star. The resonance would have the effect of greatly prolonging the unstable beryllium nucleus, giving a third helium nucleus a decent chance of hitting it. The way would then lie open to forming abundant carbon, against the apparent odds. Hoyle calculated what the energy of the resonance should be.

This was in 1951. Very little was known about the excitation of nuclei, although an experimental programme had been developed in the Second World War for the purposes of the Manhattan atomic bomb project. Hoyle was visiting Caltech at the time, and he confronted a group of American nuclear physicists, including Willy Fowler (who would go on to win a Nobel prize for related work), with his prediction of a carbon nuclear resonance. The physicists were incredulous that a little-known British astronomer could breeze in unannounced and claim to know more about carbon nuclei than a leading group of US nuclear experts. But Hoyle pestered Fowler's colleagues so persistently that they agreed to undertake an experiment to check his idea. After some modifications to their equipment, the nuclear physicists were able to announce that, indeed, Hoyle's guess was spot on. There *is* a resonance in carbon, and at just the right energy for stars to manufacture abundant quantities of this element by the triple-helium process. The experiments confirm that the resonance will prolong the lifetime of the unstable beryllium nucleus to something approaching a hundred billion-billionths of a second – long enough for the triple-helium reaction to proceed. And once the carbon is made, the rest is plain sailing. There are no more bottlenecks. Oxygen forms next, then neon, then magnesium, and so on up the periodic table of elements as far as iron. That pretty much covers all the stuff life needs to get going. Elements heavier than iron are also produced by stars, but only during explosive outbursts, when more energy is available.[9]

The carbon story left a deep impression on Hoyle. He realized that if it weren't for the coincidence that a nuclear resonance exists at just the right energy, there would be next to no carbon in the universe, and probably no life. The energy at which the carbon resonance occurs is determined by the interplay between the strong nuclear force and the electromagnetic force. If the strong force were slightly stronger or slightly weaker (by maybe as little as 1 per cent),[10] then the binding energies of the nuclei would change and the arithmetic of the resonance wouldn't add up; the universe might very well be devoid of life and go unobserved.

What are we to make of this? When Hoyle drew attention to this issue, the orthodox view was that the strength of the nuclear force is simply 'given' – it is a 'free parameter', the value of which is not determined by any theory, but must be measured by experiment. A common response was to shrug the matter aside with the comment, 'The value it has is the value it has, and if it had been different we wouldn't be here to worry about it.' But that attitude seems a bit unsatisfactory. We can certainly *imagine* a universe in which the form of the strong force law is the same but the actual strength of the force is different, just as we can imagine a world in which gravity is a little stronger or weaker, but otherwise obeys the same laws. The fact that the value of the strong and electromagnetic forces in atomic nuclei are 'just right' for life (like Goldilocks' porridge) cries out for explanation.

George Gamow, who was responsible for putting the hot big bang model of the universe on the scientific map in the early 1950s, composed the following witty account of the significance of Hoyle's discovery, which he called 'New Genesis':

In the beginning God created Radiation and Ylem.[11] And the Ylem was without shape or number, and the nucleons were rushing madly upon the face of the deep.

And God said: 'Let there be mass two.' And there was mass two. And God saw deuterium, and it was good.

And God said: 'Let there be mass three.' And there was mass three. And God saw tritium,[12] and it was good.

And God continued to call numbers until He came to the transuranium elements. But when He looked back on his work, He saw that it was not good. In the excitement of counting, He had missed calling for mass five,[13] and so, naturally, no heavier elements could have been formed.

God was very disappointed by that slip and wanted to contract the universe again and start everything from the beginning. But that would be much too simple. Instead, being Almighty, God decided to make heavy elements in the most impossible way.

And so God said: 'Let there be Hoyle.' And there was Hoyle. And God saw Hoyle and told him to make heavy elements in any way he pleased.

And so Hoyle decided to make heavy elements in stars, and to spread them around by means of supernova explosions. But in doing so, he had to obtain the same abundance curve which would have resulted from nucleosynthesis in ylem, if God would not have forgotten to call for mass five.

And so, with the help of God, Hoyle made heavy elements in this way, but it was so complicated that nowadays neither Hoyle, nor God, nor anybody else can figure out exactly how it was done.

Amen.[14]

The weak force – another 'put-up job'?

Of course, that isn't the end of the story. But before taking it further I want to run through some other 'coincidences' of a similar nature. Hoyle's 'put-up job' turned out to be the first of many instances in which seemingly small changes in some basic parameters of physics would prove lethal. A good way to think about this is to imagine playing God and setting out to design a universe. Suppose you had already settled on the basic laws of physics but you still had some free parameters at your disposal. The values of these parameters could be set by twiddling the knobs of a Designer Machine (see Figure 23). Turn one knob and the electron gets a bit heavier, turn another and the strong force becomes a bit weaker, and so on. You could do this and see what happened to the universe. When would it make a big difference, and when would it scarcely matter? Although physicists can't actually carry out the experiment (at least not yet!), they can perform simple calculations to see what – all else being equal – such changes would do to the prospects for life. The qualification 'all else being equal' is important here, because we have no idea whether the

23. *The Cosmic Designer Machine*

By twiddling the knobs on the imaginary machine, the Cosmic Designer can alter the parameters of the physical universe, such as the masses of particles and the strengths of forces. Calculations suggest that even small changes in some key parameters would wreck the familiar structure of the universe and prevent life from arising.

various parameters of interest are actually free and independent, or whether they will turn out to be linked by a more comprehensive theory, or possibly even determined completely by such a theory. Maybe you can't raise the mass of the electron *and* lower the strength of the strong nuclear force together because these two properties of nature are connected in some deep way that forbids it. From our present knowledge, however, that wouldn't seem to be the case.

So let's play with the Designer Machine and see what happens. I have discussed the strong nuclear force. What about the weak one – the one responsible for such things as radioactive decay and the transmutation of neutrons into protons? The situation here is less critical, but still interesting. The weak force is implicated in the carbon story, not only in manufacturing the carbon but in disseminating it. The carbon atoms in your body were forged inside a star somewhere, billions of years ago. How did they end up on Earth? A good way for a star to divest itself of carbon is by exploding. Massive stars typically end their lives catastrophically as supernovas. What happens is that the core of the star runs out of nuclear fuel and can no longer sustain the enormous pressure needed to hold it up against the weight of its material. A critical juncture is reached at which the core abruptly gives up and implodes catastrophically to form either a black hole or a neutron star (depending on its initial mass). The overlying material plunges inwards, following the collapsing core, but rebounds and explodes spectacularly, spewing gas into interstellar space. Stellar cataclysms like this erupt on average two or three times per century per galaxy, and release so much energy that for a few days the stricken star can rival an entire galaxy in its brightness.

A major factor in the rebound mechanism involves the weak nuclear force. When the core of a massive star implodes, its protons and electrons are violently compressed together. Under the action of the weak force, the protons turn into neutrons, and each sacrificed proton releases a neutrino in the process.[15] Thus a vast tide of neutrinos suddenly belches from the imploding core. This is not just theory: in 1987 an underground experiment in Japan devised to look for proton decay picked up a pulse of neutrinos at the same time as a supernova was seen to go off in the Large Magellanic Cloud. To the extent that neutrinos interact at all with ordinary matter, it is via the weak nuclear force. In normal circumstances this interaction is too feeble to make much difference, but the circumstances inside an imploding star are anything but normal. The nuclear material of the star collapses to a density of almost a billion tonnes per cubic centimetre – so dense that even neutrinos have a tough job ploughing through it. As they stream away from the stellar interior, they exert a powerful outward pressure, and this helps to turn around the material plunging onto the collapsing core, blasting it back into space. If the weak force were weaker, the neutrinos would lack the punch to create this explosion. If it were stronger, the neutrinos would react more vigorously with the stellar core and wouldn't escape to deliver their blow to the outer layers. Either way, the dissemination of carbon and other heavy elements needed for life via this process would be compromised.[16]

The weak force in the early universe

The weak force is important in another aspect of the life story, by controlling the amount of helium synthesized in the hot early universe. In Chapter 3, I explained how the relative abundances of hydrogen and helium depend on the ratio of neutrons to protons in the primordial material about one second after the big bang. This is how the weak force affects things. An isolated neutron is unstable, with a half-life of 615 seconds, and decays into a proton.[17] This decay process results from the action of the weak force. But it took about 100 seconds for the universe to cool enough to allow deuterium to start forming, so this was a rather close-run thing. Had the weak force been a bit stronger, the primordial neutrons would have decayed faster, reducing the total amount of helium produced, which in turn would reduce the quantity of life-encouraging carbon production in stars. On the other hand, if the weak force had been a bit weaker, a different problem would have arisen. The primordial cosmic material was a thorough mix of protons, neutrons, electrons and neutrinos. Before about one second, the different particle species were maintained at a common temperature (i.e. they were in thermodynamic equilibrium) by various interactions. Neutrinos played the determining role in maintaining the equilibrium between protons and neutrons, because these particles can transmute into each other by emitting and absorbing neutrinos (and antineutrinos). However, the ability of the neutrinos to distribute the heat energy democratically between protons and neutrons hinges on whether the transmutations take place fast enough to keep pace with the rapidly falling temperature. The race to keep up gets harder and harder, because the expansion of the universe dilutes both the energies and the densities of the participating particles, so reducing the reaction rate. Eventually the time comes when the struggle is lost. The poor neutrinos, armed only with the weak force, can't keep up, and they rather abruptly drop out of the game. This 'decoupling' event happens at a little less than one second. At that point thermodynamic equilibrium between protons and neutrons is lost, because there is no longer a mechanism to redistribute the available energy between their relative populations.

Neutrons are about 0.1 per cent heavier than protons, so the principle of democracy means that if they are to be limited to their fair share of the available heat energy, there will be fewer of them than there are protons (because it takes more energy to make a slightly heavier neutron than a proton). The degree to which this mass difference between neutrons and protons translates into a numerical advantage for the protons depends very critically on the temperature. A microsecond after the big bang, when the temperature was a trillion degrees, the 0.1 per cent mass difference was almost irrelevant (compared with the huge thermal energy available), so the ratio of neutrons to protons was nearly one to one. But as the temperature plummeted, and less and less thermal energy was available to share out, the balance of advantage favouring the lighter protons grew dramatically: the ratio of neutrons to protons sank inexorably from nearly one to one down to about six to one (six protons for every neutron). At this stage the neutrinos opted out, and the proton-to-neutron ratio remained stuck at a little more than six to one.

So now we can see what would have happened if the weak force had been weaker. The neutrinos would have given up the struggle sooner, when the universe was hotter, and when the numerical advantage given to the lighter protons by the democracy principle was less. That would have meant more neutrons and fewer protons in the final mix. Because the excess protons went on to make hydrogen, there would have been less hydrogen in the universe and more helium. Had the freeze-out ratio been exactly one to one, *all* of the material would have ended up as helium. Less hydrogen would have had dire consequences for life. Stable, long-lived stars like our sun are hydrogen reactors. Without an abundant supply of this raw material they would be starved of fuel and would have very different properties. Also, hydrogen combines with oxygen to make water, a vital part of the life story at all stages. For example, life probably began in a watery 'primeval soup', and for the greater part of its history life on Earth has been confined to the oceans. Even land animals like us contain about 75 per cent water. Without abundant water, the chances of life forming and flourishing are slim.

The upshot of these various nuclear considerations, then, is that had the weak force been either somewhat stronger or very slightly weaker, the chemical make-up of the universe would be very different, and with much poorer prospects for life.

Fine tuning in the other forces

Let me now turn to the other two forces of nature, gravitation and electromagnetism. How vital are their properties to the life story? It is easy to see why changing their strengths too much would threaten biology. If gravity were stronger, stars would burn faster and die younger: if by some magic we could make gravitation twice as strong, say, then the sun would shine more than a hundred times as brightly. Its lifetime as a stable star would fall from 10 billion to less than 100 million years, which is probably too short for life to emerge, and certainly too short for intelligent observers to evolve. If electromagnetism were stronger, the electrical repulsion between protons would be greater, threatening the stability of atomic nuclei.

One of the striking things about electromagnetism and gravitation is the vast disparity in their relative strengths. In a normal hydrogen atom, the single electron is bound to the single proton by electrical attraction. But another source of attraction is at work here too – gravitation. It is easy to calculate the relative strengths of these two forms of attraction. It turns out that the electrical force is about 10^{40} times stronger than the gravitational force. Clearly, then, gravity is extraordinarily weak compared with electromagnetism. That isn't how we experience these forces, though. We feel Earth's gravity a lot, whereas everyday electrical forces seem insignificant by comparison. The reason for this disparity is because gravitation is a cumulative effect: the more matter there is, the stronger is its gravity. The situation with electric charges is different, because they come in both negative and positive varieties. Accumulate a lot of positive charge somewhere, and it will attract negative charge from its environment, thus reducing the net force. In this way electric charge has a natural self-limiting property. Not so with gravitation: the more matter there is in one place, the more it drags in additional material and the stronger the combined gravitational force becomes. Gravitation is therefore inherently self-amplifying, so that in spite of its extreme weakness it can add up to become dominant, as happens when a star collapses.

Years ago, Brandon Carter found an amazing relationship between the unexplained ratio 10^{40} and the properties of stars. Every star must transport heat from the nuclear furnace in its core to the surface, where it radiates into space. Heat can flow in two ways: by radiation, in which photons convey the energy, and by convection, in which hot gas from deep down rises to the surface, bringing heat with it. Our sun has a convective outer layer, and through a telescope its surface looks like a boiling maelstrom. Astronomers think that this convective motion plays a role in forming planets, although it is far from clear how (the process of planet formation is still poorly understood). Larger stars rely on radiative transfer of heat rather than convection, and this is thought to be important in creating the conditions that lead to supernova explosions. Because both planets and supernovas are a major part of the life story, it is important for the universe to contain a selection of both radiative and convective stars. Carter discovered from the theory of stellar structure that to get both sorts of stars, the ratio of the strengths of the electromagnetic and gravitational forces needs to be very close to the observed value of 10^{40}. If gravity were a bit stronger, all stars would be radiative and planets might not form; if gravity were somewhat weaker, all stars would be convective and supernovas might never happen. Either way, the prospects for life would be diminished.

More fine-tuning marvels

As if what I have already described isn't enough, there are many more 'convenient coincidences' in basic physics that go to make the universe bio-friendly. Other examples are the masses of the various subatomic particles. Physicists produce tables of these, and the numbers are fascinating but on the face of it almost totally meaningless. I receive lots of unsolicited manuscripts from mystically minded amateur scientists convinced that they have spotted subtle patterns in the numerical values of these masses. Sadly, such schemes are all bogus. Maybe one day theorists will be able to derive these numbers from some deep mathematical principles tied to a proper physical theory, but that is a distant prospect. Meanwhile, we can take the numbers as given, stare at them, and ask what they imply for life.

To give you a feeling for what I am talking about, the ratio of the mass of the proton to that of the electron is 1,836.152 667 5 – an utterly mundane number. The neutron-to-proton mass ratio is 1.001 378 418 70, which looks equally uninspiring. Physically, it means that the proton has very nearly the same mass as the neutron, which, as we have already seen, is about 0.1 per cent heavier. Is this important? Indeed it is, and not just in determining the ratio of hydrogen to helium in the universe. The fact that the neutron's mass is coincidentally just a little bit more than the combined mass of the proton, electron and neutrino is what enables free neutrons to decay. If the neutron were even *very* slightly lighter, it could not decay without an energy input of some sort. If the neutron were lighter still, yet only by a fraction of 1 per cent, it would have less mass than the proton, and the tables would be turned: it would be isolated protons rather than neutrons that would be unstable. Then protons would decay into neutrons and positrons, with disastrous consequences for life, because without protons there could be no atoms, and no chemistry.

Cosmology provides more remarkable examples of fine tuning. As I have discussed, the cosmic microwave background radiation is embellished with the all-important ripples or perturbations, echoes of the seeds for the large-scale structure of the universe. These seeds, remember, are thought to originate in quantum fluctuations during inflation. Numerically, the variations are small: about one part in a hundred thousand, a quantity which cosmologists denote by the letter Q. Now, if Q were smaller than one-hundred-thousandth – say one-millionth – this would severely affect the formation of galaxies and stars. Conversely, if Q were bigger – one part in ten thousand or more – galaxies would be denser, leading to lots of planet-disrupting stellar collisions. Make Q too big and you'd form giant black holes rather than clusters of stars. Either way, Q needs to sit in a rather narrow range to make possible the formation of abundant, stable, long-lived stars accompanied by planetary systems of the type we inhabit.

Returning to my Designer Machine metaphor, the collection of felicitous 'coincidences' in physics and cosmology implies that the Great Designer had better set the knobs carefully, or the universe would be a very inhospitable place. How many knobs are there? The Standard Model of particle physics has about twenty undetermined parameters, while cosmology has about ten. All told, there are over thirty 'knobs'.[18] As I have cautioned already, not all the parameters are necessarily independent of the others, and not all require exceptional fine tuning for life to be possible. But several certainly do: some of the examples I have given demand 'knob settings' that must be fine-tuned to an accuracy of less than 1 per cent to make a universe fit for life. But even this sensitivity pales into insignificance compared with the biggest fine-tuning riddle of all: dark energy.

The biggest fix in the universe

I have described dark energy as a cosmic repulsion, or antigravity force, which drives the galaxies apart at an accelerating rate. But actually this is a bit misleading because the antigravity still acts when no ordinary matter is present at all. As I mentioned briefly in Chapter 3, if the cosmic repulsion force is included, then a completely empty universe will expand 'exponentially,' doubling in volume in successive fixed intervals of time. To recap, one can think of empty space as being filled with invisible dark energy and its associated negative pressure, a combination which creates antigravity.

Why should empty space possess dark energy? Why isn't it simply empty, with no energy whatsoever? One reason, which I alluded to in Chapter 3, is that space would possess dark energy if it were permeated by an invisible scalar field, such as the inflaton field. We wouldn't be able to see or touch this field, but it would still generate antigravity, as it is supposed to do with a vengeance during the inflationary phase of the very early universe. But there is another reason too, provided by quantum mechanics, which predicts that even apparently empty space is teeming with virtual particles (see Box 4, p. 72). Virtual particles, like real particles, possess energy, and it turns out that they also possess exactly the requisite negative pressure to generate a cosmic repulsion force of the sort that Einstein proposed.

I have already pointed out that when Einstein introduced the cosmic repulsive force 'by hand' into his general theory of relativity, the theory itself couldn't pin down its value. He was free to choose any number he wanted to multiply the antigravity term and thereby set the overall strength of the cosmic repulsion force. In the event, he used astronomical data to calculate a plausible value that would permit a static universe – his favoured model. When he later changed his mind about a static universe, it was just a simple matter of adjusting the number multiplying the cosmic repulsion term to zero, thus eliminating that term from the equations altogether. Such a fast and loose approach to cosmic repulsion, or dark energy, might be justified if one is restricting the analysis to gravitation only, but when quantum mechanics comes in, it simply won't do.

About thirty years ago, several theoretical physicists, myself included, decided to work out how much dark energy would be provided by all the virtual particles that inhabit the quantum vacuum (see Box 7 for more details). One can consider, say, the electromagnetic field and determine how much quantum energy resides in a given volume of 'empty' space (i.e. with no 'real' photons present). The calculation isn't hard: a rough answer can be worked out on the back of an envelope. The answer, unfortunately, is unbelievable. Converted into mass density, it comes out at about 10^{93} grams per cubic centimetre, implying that a thimbleful of empty space should contain a million trillion trillion trillion trillion trillion trillion tillion tonnes! Stephen Hawking once quipped that this must be the biggest failure of physical theory in history. How could we get it so wrong?[19]

Faced with such a serious embarrassment, physicists scrambled for an explanation. Perhaps some sort of cancellation mechanism might be at work. The electromagnetic field is only one of many fields in nature, and some of the others contribute *negative* dark energy. Perhaps there is a deep symmetry that causes the pluses and minuses to neutralize each other exactly. And in fact such a symmetry exists: it is supersymmetry. The trouble is, we know that supersymmetry is broken in the real world, and unless the symmetry is exact, the positives and negatives won't cancel out. Many other ideas were tried, but they all looked contrived. Nevertheless, it was possible to believe –

7. *Dark energy and the quantum vacuum*

Quantum mechanics predicts that even empty space is replete with invisible energy. Here's why. Imagine a pendulum consisting of a ball suspended from a string. When the pendulum is swinging, it possesses two forms of energy: kinetic energy (energy of motion) and potential energy (energy gained by the ball as it rises above the lowest point of its swing). The pendulum has zero energy when the string is vertical and the ball is hanging at rest at the lowest point.

Quantum mechanics changes this simple picture. Heisenberg's uncertainty principle forbids the ball from simultaneously having both a well-defined motion and a well-defined location: there is a trade-off in uncertainty between them (see Box 4, p. 72). If the ball is well located near the lowest point, thus reducing its potential energy almost to zero, it will have a large uncertainty in its motion, and so cannot possess zero kinetic energy. Conversely, if the ball is considered to be almost at rest – that is, with near-zero kinetic energy – it will have a large uncertainty in its vertical position, and so will possess potential energy. A careful calculation shows these two quantum contributions to energy always add to the same result, called the *zero-point energy* of the pendulum. Its average, or expectation, value depends on the natural frequency of the pendulum's oscillations: the faster the frequency of oscillation, the bigger the zero-point energy.

All quantum systems that can oscillate, for example atoms in a crystal lattice, or diatomic molecules, have an irreducible zero-point energy. Even electromagnetic waves have zero-point energy. This is no surprise: wave motion is oscillatory. Electromagnetic waves can have any wavelength, and each wavelength possesses its own irreducible zero-point energy, which exists even if no photons are present. The shorter the wavelength, the greater the frequency of the wave and the larger the zero-point energy associated with it.

A simple calculation summing up the zero-point energy associated with *all* possible wavelengths is easy to perform. A decision has to be made where to stop this summation, because there are an infinite number of wavelengths (the electromagnetic field can be represented as an infinite collection of oscillators), and infinitesimal wavelengths have infinite energy. If you just add up the lot you get the answer that the vacuum contains *infinite* quantum zero-point energy. An appropriate place to truncate the sum is at the Planck length (see p. 85), because the zero-point energy at that frequency is so large that it starts to bend space into weird shapes. Stopping there yields a natural value for the density of dark energy – natural because no quantities have been inserted into the theory other than the constants of nature, G, h and c, which come into the theory anyway. Expressed as a mass density, the value obtained from such a summation is about 10^{93} grams per cubic centimetre, a truly stupendous overestimate compared with the measured value of dark energy – a paltry 10^{-28} grams per cubic centimetre.

and *was* believed by most physicists and cosmologists – that *some* physical mechanism drove the value of dark energy (or cosmic repulsion) to precisely zero.[20]

Such hopes were thoroughly dashed when astronomers discovered that dark energy is actually *not* zero after all. This came as a complete shock. The value of the dark energy mass density measured by astronomers is some 120 powers of ten *less* than the 'natural' value obtained by applying quantum theory to the virtual particles in a vacuum (see Box 7). When the value of dark energy seemed to be zero, it was at least plausible that some yet to be discovered mechanism might operate to force an exact cancellation. But, as Leonard Susskind has stressed,[21] a mechanism which cancels to one part in 120 powers of ten, and then fails to cancel after that, is something else entirely. To give the reader some idea of just how much of a fix this almost-cancellation is, let me write out the number 10^{120} in its full glory:

1,000,000,000,000,000,000,000,000,000,000,000,000,000,
000,000,000,000,000,000,000,000,000,000,000,000,000,
000,000,000,000,000,000,000,000,000,000,000.

So the big fix[22] somehow works brilliantly (if mysteriously) for 119 powers of ten, but *fails* at the 120th.

Whatever dark energy may be – and it may just be the 'natural' energy of empty space – it is dangerous. In fact, it could be the most dangerous stuff known to science. About twenty years ago Steven Weinberg pointed out that if the magnitude of the dark energy were only moderately larger than the observed value, it would have frustrated the formation of galaxies.[23] Galaxies form by the slow aggregation of matter under the action of attractive gravitation. If this tendency were opposed by a strong enough cosmic repulsion force, galaxies would be unable to grow properly. And as I have already remarked, without galaxies there would probably be no stars or planets or life. So our existence depends on the dark energy not being too large. A factor of ten would suffice to preclude life: if space contained ten times as much dark energy as it actually does, the universe would fly apart too fast for galaxies to form. A factor of ten may seem like a wide margin, but one power of ten on a scale of 120 is a pretty close call. The cliché that 'life is balanced on a knife-edge' is a staggering understatement in this case: no knife in the universe could have an edge *that* fine.[24]

Logically, it is possible that the laws of physics conspire to create an almost but not quite perfect cancellation. But then it would be an extraordinary coincidence that *that* level of cancellation – 119 powers of ten, after all – just happened by chance to be what is needed to bring about a universe fit for life. How much chance can we buy in scientific explanation? One measure of what is involved can be given in terms of coin flipping: odds of 10^{120} to one is like getting heads no fewer than *400 times in a row*. If the existence of life in the universe is completely independent of the big fix mechanism – if it's just a coincidence – then those are the odds against us being here. That level of flukiness seems too much to swallow.

But what is the alternative? There is indeed another way to explain the minuscule value of the dark energy, and perhaps all the other convenient 'coincidences' in physics and cosmology, but it represents a huge departure from the way we normally do science, and many scientists are aghast at it. But, as we shall see in the next chapter, it may be the only answer.

Key points

- The existence of life as we know it depends delicately on many seemingly fortuitous features of the laws of physics and the structure of the universe.
- An early famous example of how the laws of physics seem to be fine-tuned for life is the production of carbon in stars, which requires a numerical 'coincidence' to produce a nuclear resonance at just the right energy.
- All four forces of nature are implicated in the life story. Changing the strength of any one of them, even by a small amount, could render the universe sterile.
- The masses of some fundamental particles could not be very different without compromising the habitability of the universe.
- The measured value of dark energy is 120 powers of ten less than its natural value, for reasons which remain completely mysterious. If it were 119 rather than 120 powers of ten less, the consequences would be lethal.

8

Does a Multiverse Solve the Goldilocks Enigma?

We might be winners in a cosmic lottery

Scientists have long been aware that the universe seems strangely suited to life, but they mostly chose to ignore it. It was an embarrassment – it looked too much like the work of a Cosmic Designer. Discussion of the anthropic principle was frowned upon as being quasi-religious. Andrei Linde says that in the old Soviet Union, only one person ever worked on it.[1] Today the mood has changed. What made a difference was the idea of a *multiverse*, which offers the opportunity to explain the weird bio-friendliness of the universe as a straightforward selection effect, without invoking divine providence.

The multiverse theory says that what we have all along been calling 'the universe' is in fact nothing of the kind. Rather, it is but an infinitesimal fragment of a much larger and more elaborate system – an ensemble of 'universes', or of distinct cosmic regions (such as the 'pocket universes' that feature in the eternal inflation theory). Imagine that these universes, or regions, differ in some property which is important for life. Then – obviously – life will arise only in those universes, or cosmic regions, where conditions favour life. Universes which cannot support life will go unobserved. It is therefore no surprise that we find ourselves located in a universe which is suited to life, for observers like us could not have emerged in a sterile universe. If the universes vary at random, then we would be winners in a gigantic cosmic lottery which created the illusion of design. Like many winners of national lotteries, we may mistakenly attribute some deep significance to our having won (being smiled on by Lady Luck, or suchlike) whereas our success really boils down to chance.

Let me give an example where this type of argument seems to work successfully. Forget about inflation for a moment and suppose that the universe began with a conventional big bang. Imagine that, instead of it being a smooth and uniform affair, its vigour varied randomly from place to place (on a very large scale). In some regions the bang would lack the power to spread the matter out: those regions would rapidly collapse into gigantic black holes. No life there. In other regions the bang would be so big that the material would disperse too rapidly for galaxies or stars to form. No life there either. But here and there, by chance alone, a Goldilocks region resembling our own cosmos would emerge in which the expansion rate was just right: slow enough to tolerate a limited amount of gravitational clustering (to form galaxies and stars) but not so slow as to suffer catastrophic collapse (into a huge black hole). We need not be astonished to find ourselves in such a well-ordered cosmic region, even if it is exceptional, and even if it arose against vast odds by mere chance – for the conditions within it are precisely the conditions conducive to life.

Thirty years ago, some cosmologists appealed to this type of anthropic selection effect as an explanation for why the cosmological expansion seems to be just right for life, in both its rate and its uniformity. However, the inflationary scenario automatically explains these life-encouraging features in terms of a physical theory, so the anthropic explanation was abandoned. But the problem isn't yet completely solved, because one still needs to assume the right level of primordial density perturbations to create galaxies, perturbations that probably arose from quantum fluctuations during the inflationary phase. Why the fluctuations in our universe have the amplitude they do is unknown. The answer might turn out to be an inevitable consequence of a future theory, or it could be that the strength of the fluctuations varies from region to region, in which case there may still be a degree of anthropic selection involved.

Cosmic structure is one thing, but can the laws of physics *vary?*

This is all well and good when it comes to the structure of the universe, but what about the other examples of fine tuning I discussed in the last chapter, such as Hoyle's famous carbon resonance coincidence – properties which seem to require rather precise values for the relative strengths of the strong and electromagnetic forces? Or of the masses of various subatomic particles? To explain these latter 'coincidences' anthropically – that is, in terms of observer selection – *the laws of physics themselves* would have to vary from one cosmic region to another. Is this credible? If so, how could it happen?

We can gain some insight into this from history. After Copernicus established that the planets go round the sun, Kepler and others tried to make sense of the numerical relationships found in the solar system. In those days just six planets were known, and their various distances from the sun had been measured to reasonable accuracy. It was natural to ask, why six planets? And why those distances? Was there some deep principle of nature, some law-like mathematical scheme, which would yield the observed numbers? Kepler came up with an ingenious idea based on ancient geometrical shapes. He envisaged the planetary orbits attached to spheres inside perfect polyhedra nested inside one another, following the mystical tradition of the Pythagoreans, who two millennia before had tried to interpret the cosmos in terms of musical and geometrical harmony. Later, in the eighteenth century, the German astronomer Johann Bode published a simple numerical formula (which became known as Bode's law) that gave the distances of all the six known planets from the sun, plus a 'missing' one, where the asteroid belt was later discovered to be located.

Today these attempts to shoehorn the solar system into a neat and elegant arithmetic pattern seem silly, and the superficial concordance of Bode's formula with the measured planetary distances is revealed to be a lucky coincidence. We now know that the arrangement of the planets is largely a historical accident. The planets formed from a swirling nebula of gas and dust which surrounded the proto-sun. There were more planets initially than the nine we know today. Some collided and merged, others were flung out of the solar system altogether. All the planets' orbits have changed somewhat over the 4.5 billion years since the solar system formed. The point is that what we have ended up with is the result of chaotic circumstance – the amount of material in the solar nebula, the complicated forces that caused the planets to congeal where they did, the disturbances of nearby stars and gas clouds. Clearly there is nothing *fundamental* about the planets and their distances: these features are purely *incidental*. We now know of other planetary systems around other stars which have completely different orbital arrangements. So what was once thought to be a deep law of nature turned out to be just a frozen accident of history, albeit an important one as far as we are concerned (if Earth hadn't acquired roughly the orbit is has, it would be uninhabitable).

Drawing on the lesson of the solar system, it makes sense to ask whether other features of the world which we currently regard as law-like might also turn out to be accidents of history. Could it be that some of the regularities of nature which we dignify as 'laws of physics' are actually frozen relics from the formation of the universe? If that were so, it would seem reasonable to suppose that other regions of the universe, or other pocket universes, possess different laws in much the same way that other star systems possess different planetary arrangements.

The 'constants' of nature, at least, might vary

Let me recap an important point. Laws of physics have two features which might in principle vary from one universe to another. First, there is the mathematical form of the law, and second, there are various 'constants' that come into the equations. Newton's inverse square law of gravitation is an example. The mathematical form relates the gravitational force between two bodies to the distance between them. But Newton's gravitational constant G also comes into the equation: it sets the actual strength of the force (see Box 6, p. 88). Another example is the Dirac equation for the electron. It too has a particular mathematical form, which describes how the electron moves in accordance with quantum mechanics and relativity. It also contains three constants: the speed of light, the mass of the electron and Planck's constant. When speculating about whether the laws of physics might be different in another cosmic region, we can imagine two possibilities. One is that the mathematical form of the law is unchanged, but one or more of the constants takes on a different value.[2] The other, more drastic possibility is that the *form* of the law is different. In this section I shall restrict discussion to the former.

The Standard Model of particle physics has twenty-odd unfixed parameters. These are key numbers such as particle masses and force strengths which cannot be predicted by the Standard Model itself, but must be measured by experiment and inserted into the theory by hand. Nobody knows whether the measured values of these parameters will one day be explained by a deeper unified theory that goes beyond the Standard Model, or whether they are genuinely free parameters which are not determined by *any* deeper-level laws. If the latter is correct, then the numbers are not God-given and fixed but could take on different values without conflicting with any physical laws. By tradition, physicists refer to these parameters as 'constants of nature' (see Box 6, p. 88) because they seem to be the same throughout the observed universe. However, we have no idea *why* they are constant and (based on our present state of knowledge) no real justification for believing that, on a scale of size much larger than the observed universe, they *are* constant. If they can take on different values, then the question arises of what determines the values they possess in our cosmic region.

A possible answer comes from big bang cosmology. According to orthodox theory, the universe was born with the values of these constants laid down once and for all, from the word go. But some physicists now suggest that perhaps the observed values were generated by some sort of complicated physical processes in the fiery turmoil of the very early universe. If this idea is generally correct, then it follows that the physical processes responsible could have generated *different* values from the ones we observe, and might indeed have generated different values in other regions of space, or in other universes. If we could magically journey from our cosmic region to another region a trillion light years beyond our horizon we might find that, say, the mass or charge of the electron was different. Only in those cosmic regions where the electron mass and charge have about the same values as they do in our region could observers emerge to discover a universe so propitiously fit for life. In this way, the intriguingly life-friendly fine tuning of the Standard Model parameters would be neatly explained as an observer selection effect.

The origin of mass, and why it might vary

When I was a student I was given a table of the masses of the various subatomic particles and simply told, 'That's what they are.' Questions about why those numbers and not others were dismissed as mystical nonsense. Today it is acceptable to demand some sort of explanation for mass. In fact, we have such an explanation: it is called the Higgs mechanism. I mentioned it in passing in Chapter 4 in connection with the theory of the electroweak force. It is a crucial part of the Standard Model. In a nutshell, the Higgs theory goes like this. Electrons and quarks have no intrinsic mass; instead, they acquire their mass by interacting with an invisible field which pervades all of space, rather like the aether of old. It is the Higgs field that gives these particles their mass. The amount of mass they end up with depends on how strongly the particles sense the Higgs field – how strongly they *couple* to it, to use the correct terminology. The photon doesn't interact with the Higgs field at all, so it remains massless. Quarks couple to the Higgs field much more strongly than electrons: the top quark feels it most, and ends up with a mass hundreds of thousands of times greater than the electron. The W and Z particles, which mediate the weak force, also acquire their very considerable mass by coupling to the Higgs field.[3]

Most theoretical physicists are persuaded that the Higgs field really exists, even though we can't directly discern it. Their hopes for clinching evidence have focused on the Large Hadron Collider (LHC), the giant accelerator machine currently being built at CERN. The plan is for the LHC to use high-energy proton–antiproton collisions to make a Higgs particle by 2010. Just as the electromagnetic field has an associated quantum particle in the guise of the photon, so the Higgs field should have a quantum particle too, called the Higgs boson.[4] But whereas the photon is massless, the Higgs boson is predicted to have an enormous mass, greater than 180 proton masses – which is why it has never been made in an experiment before. (The Higgs particle acquires its own very considerable mass by interacting with its own field.)

Meanwhile, back at the theorist's desk, I still need to explain how the masses of particles might vary from one cosmic region to another. One obvious way is if the strength of the Higgs field changes from one to the other. There are reasons why this could possibly happen, and numerous mathematical models have been published which link the Higgs field with other aspects of physics to yield such variability, but I don't want to get into the technicalities here. I just want to establish the general point that if the Higgs field did vary from one place to another, then the mass of the electron and the masses of the quarks would vary as well. Over here the electron would have mass m, over there it would have a different mass, m^1. What's more, the ratio of the mass of the proton to the mass of the electron would vary too. Although quark and electron masses would vary in lockstep, a proton is an assemblage of three whirling quarks – a little bundle which contains a lot of kinetic, electrical and gluon energy too. In fact, most of the total mass of the proton is in the form of this additional energy. And this contribution would *not* track the electron and quark mass variations. So if, for example, the Higgs field doubled in strength, the electron's mass would double, but the proton would become only modestly heavier. Crucially, then, the *ratio* of the electron to proton mass would change, and if it changed enough it would throw out the delicate life-encouraging fine tuning, for example by drastically changing nuclear reactions in the early universe and in stars.

Symmetry-breaking explains how simple laws can produce a complicated world

There is a very general reason why variations in some of the basic parameters, or 'constants', of physics might be expected. It is called *symmetry-breaking*. Here is a simple example. An icon of modern science is the double helix – the spiral figure of the life molecule DNA, made famous by James Watson's bestselling book.[5] The DNA in your cells contains the genetic blueprint – the genome – that makes you *you*. All DNA molecules, not just in humans but in all known life forms, are coiled in the form of a helix. More specifically, they have the configuration of a right-handed helix. Spiral staircases in ancient castles sometimes wind to the left as you go up them, and sometimes to the right. DNA always winds to the right.

There is no fundamental reason why life couldn't use left-handed DNA. It would be chemically identical, equally stable, and nothing about it would violate any laws of physics. That is because the laws of electromagnetism – which are responsible for building molecules – are completely indifferent to the distinction between left and right. Expressed more carefully, physicists say that electromagnetism has mirror symmetry. To be sure, life would get into a muddle if it tried using both left and right forms together, but there is no particular reason why right is right. The best guess is that, long ago when life was forming (somehow) from non-life, a random molecular accident broke the symmetry, and once an arbitrary choice had been made it got frozen in. It had to be frozen in so that all life could use a common standard, but there is a 50 per cent chance that it could have gone the other way.[6]

Like DNA, many physical systems do not explicitly manifest the underlying symmetries of the forces that shape them. It's not hard to find examples. Earth goes around the sun anticlockwise when viewed from above the northern hemisphere, but Newton's laws of motion and gravitation care nothing for clockwise or anticlockwise – they are symmetric. If Earth orbited in the other direction, the laws of physics would be perfectly happy.

But other physical systems do respect the underlying symmetry, at

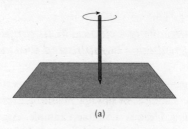

(a)

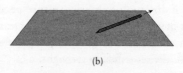

(b)

24. *Spontaneous symmetry-breaking*

(a) A pencil poised on its tip is in a symmetric state: it looks the same if rotated about a vertical axis, as shown. In this state the pencil manifests the underlying symmetry of the gravitational field, which is indifferent to horizontal directions. However, the symmetric state is unstable, and the pencil will topple over. (b) In the horizontal state, the pencil is stable but it has broken the symmetry of the gravitational field by selecting a specific horizontal direction, indicated by the arrow. The actual direction is arbitrary. If the experiment were performed many times, the pencil's positions would distribute themselves randomly around the toppling point: the ensemble would manifest the underlying symmetry, even though each specific case breaks it.

least approximately. For example, the sun is almost exactly round, reflecting the fact that the force of gravitation does not distinguish one direction in space from another: a more precise way to say this is that gravitation is symmetric under rotation. Whether or not the state of a physical system respects or breaks underlying symmetries in the laws of physics turns on questions of stability. It may be that a broken symmetry is a more stable state. A good illustration is shown in Figure 24. Imagine holding a sharp pencil exactly vertically with its tip on a horizontal surface. When you let go, the pencil topples over and falls flat. There is no profound significance in the final orientation that the pencil adopts: it is random.[7] Do it a thousand times and you will get a thousand different directions, distributed more or less evenly around the centre. The reason is that Earth's vertical gravitational field doesn't care about horizontal directions; one is the same as another as far as it is concerned. Stated more precisely, the gravitational field is symmetric under rotations around the vertical axis defined by the pencil's initial upright position. But the final state of the pencil, lying flat, breaks that rotational symmetry by selecting a *particular* horizontal direction. The state that conforms to the underlying symmetry of the laws of physics (vertical pencil) is unstable, so the pencil breaks the symmetry and topples over to a state (horizontal pencil) which is stable, but does not manifest the symmetry. So the pencil trades symmetry for stability. The symmetry is still there in the underlying laws, but you cannot discern it from an individual case. Only by inspecting a large collection of states (thousands of pencils) spread evenly over the available horizontal directions would the underlying rotational symmetry be manifest. When there is a trade-off between symmetry and stability like this, it is called *spontaneous symmetry-breaking*, because the system itself chooses (arbitrarily) how to break the symmetry – it is not imposed by some external influence.

Symmetries, and simplicity, can be restored by intense heat

Now we come to a crucial point. Broken symmetries can often be 'restored' (i.e. made manifest) by raising the temperature. Think of the DNA molecule. Above about 100°C its stability starts to be threatened by thermal agitation: it begins to melt. Heat DNA to 200°C and it falls apart completely. Its helical structure vanishes. The components fly about chaotically, and all semblance of right-handedness disappears. The chaotically moving and rebounding components faithfully display the underlying left–right symmetry of electromagnetism in a way that the assembled DNA molecule didn't.

The symmetry-breaking literally melts away with rising temperature. The general rule is this:

High-temperature systems manifest more symmetry than low temperature systems. When the temperature falls, symmetries get broken.

We shall shortly see how important this rule is for understanding the nature of the universe, but first I'd like to illustrate it with an example closer to high-school physics, as it is something of a classic. It was first enunciated by Pierre Curie, husband of the renowned chemist Marie Curie, and it concerns magnets. It was long known that the magnetic field of, say, an iron bar gets weaker when the bar is heated. A critical temperature is reached, called the Curie temperature, at which the magnetic field outside the bar vanishes (for iron, this is 770°C). It is easy to understand why. Iron atoms possess magnetic fields due to their whirling electrons, and for reasons connected with Pauli's exclusion principle the atoms like to line up their fields in parallel to form microscopic magnetic domains. In a bar magnet the domains are also aligned, thereby combining their magnetic fields in an orderly way. When the temperature is raised, however, the mini-magnets start to jostle about, trying to break free of the magnetic regimentation. The balance of advantage in this tussle tips more and more in favour of freedom as the temperature rises and the resulting agitation gets more frenetic. Eventually a critical point is reached where anarchy sets in and all the mini-magnets become independent. Under these circumstances their orientations are random: there is no systematic line-up in any particular direction. So although the hot iron is made of magnetized atoms, the net magnetic field averages to zero. If the iron is now *slowly* cooled back down again through the Curie temperature, magnetization can return. It takes its lead from any external magnetic field that may be present, such as the Earth's field, and the domains line up in that direction. If it is cooled quickly, however, the domains get frozen in their random orientations. Within each domain the magnetism has a consistent direction, but the bar, although made of magnetic material, has no overall magnetization.

Symmetry enters into this story in an obvious way. The laws of electromagnetism are symmetric under rotation (i.e. they are indifferent to direction in space). The magnetic state of the hot iron conforms to this symmetry: it points nowhere in particular because it has zero net field strength. But below the Curie temperature, magnetic fields spontaneously set in as each mini-magnet locks into a particular direction. In so doing, each one breaks the rotational symmetry of the underlying laws of electromagnetism, which control the behaviour of the system. If the bar has been cooled quickly, freezing the randomly oriented domain structure, then on a larger-scale view the rotational symmetry is still apparent because the bar is not systematically magnetized in any particular direction.

You can't tell by looking at it whether an iron bar is magnetized or not, but another example, in which the symmetry-breaking is obvious at a glance, is when water freezes to ice. The solid phase congeals from the liquid phase as the temperature falls through 0°C. Again, rotational symmetry is broken: liquid water is the same in all directions, whereas ice crystals form regular geometrical shapes with definite orientations. Physicists refer to an abrupt change of this sort as a *phase transition*.

A useful way to think about symmetry is in terms of structure and complexity. The more symmetry a system possesses, the simpler and less structured it is: compare, for example, the symmetric figure of a circle with an irregular polygon. Raising the temperature of a physical system reduces structure and destroys complexity: think how much simpler a glass of water is than a tumbler full of ice cubes. Or imagine putting Earth in a colossal furnace and turning up the heat. First the glaciers and icebergs melt, then the forests burn and the oceans boil; eventually the mountains turn molten. With enough heat, the entire planet would vaporize. A giant furnace full of vapour is much simpler – it has more symmetry – than the complex structure of Earth. The general principle is: heat = blandness, cold = richness.

Symmetry-breaking just after the big bang

Now let me return to cosmology. No furnace in the universe compares to the hot big bang. The very early universe was so hot that everything we know was literally in the melting pot (well, the vaporization pot). All the richness, diversity and complexity of the universe we now see lay in the future: uniformity was the order of the day. So the universe began with a high degree of symmetry, but as it cooled, more and more symmetries became broken, and more and more structure and complexity emerged, much of it random and spontaneous. These transformations took place through a succession of phase transitions. The reason we can make any progress at all with a theory of the early universe is precisely because it was so simple, and its pristine simplicity stemmed from the enormous temperatures and pervasive symmetry of the primordial phase.

The general trend from simple to complex, from blandness to rich-ness, as the universe cooled from its fiery birth is easy enough to understand. But we need to go further than this and address the following question: Could the spontaneous appearance of structure in the early universe extend to the 'constants of nature'? Might the values of the undetermined parameters of the Standard Model, such as particle masses and force strengths, also be randomly frozen accidents resulting from symmetry-breaking phase transitions?

Let's return to the example of the magnet. Imagine a tiny man located deep within the iron magnet, inside one of the domains.[8] He is surrounded by a magnetic field pointing consistently in one direc-tion. This pervasive magnetic field is part of the little man's universe, and the behaviour of electric charges and magnets in his vicinity are all affected by this field. If the man worked out the laws of electromagnetism within his tiny universe, he would be obliged to incorporate into them this ambient field. Obviously these laws would not be symmetric under rotation because the field has a particular direction. With our god's-eye view we can see that the little man is misled. What he takes as a basic law of his universe, a law which breaks rotational symmetry, we can see is just a frozen accident, the specifics of which are limited to that particular domain.

What if the iron bar had no overall magnetization, but a randomly oriented domain structure? Then the little man could, if he travelled far enough, cross into a neighbouring domain. He would then get a shock because the magnetic field in this 'universe next door' would point in a different direction, and he would need a different set of laws. If he visited enough domains, and experienced the random orientation of many local fields, he might become convinced that his laws of electromagnetism were not the true, fundamental laws, but involved a frozen accident (the field direction) which broke an important underlying symmetry (rotational symmetry). He would conclude that what he took to be a fundamental law is in fact merely an 'effective' law, beneath which lies a truly fundamental electromagnetic law which is rotationally symmetric. If he knew enough physics he might also be able to work out that if his universe and its neighbouring domains were heated above the Curie temperature, this symmetry-breaking would melt away, the domains would merge, and the full symmetry of the true, underlying law of electromagnetism would reveal itself everywhere.

This story is an effective metaphor for early-universe cosmology. It suggests that when a familiar law of physics involves the breaking of some symmetry, then the law might actually turn out to be a low-temperature, or low-energy, *effective* law, valid in our particular cosmic domain but randomly different in other cosmic domains. Like the little man in the magnet, we too might have a 'universe next door' where the symmetry is broken differently, and the low-energy physics that stems from it is also different. And back in the ultra-hot primeval phase which followed the big bang, the symmetry would have been manifest everywhere, and the cosmic domains would have been indistinguishable. A domain structure of some sort can be expected in almost any cosmological model in which there is cooling from a hot big bang plus symmetry-breaking. If the symmetry-breaking determines one or more of the parameters that need to be fine-tuned for life to arise, then we have a ready explanation for the Goldilocks enigma. Only in domains where the effective, low-energy, low-temperature laws come out favourably for life, purely by accident, will observers be possible.[9]

Abstract symmetries are crucial in physics

It is easy enough to understand the experience of the little man in the magnet, because the symmetry in that example – rotational symmetry – is familiar in daily life. But the sort of symmetry that might lead to a cosmic domain structure is different. It is, in fact, an 'abstract' symmetry, not a geometrical one. Let me explain this by giving an example from economics. It concerns inflation, but of the financial as opposed to the cosmological variety. Inflation notoriously erodes the purchasing power of money, but the *intrinsic* value of goods and services is not affected. In the days of pegged exchange rates, governments sometimes abruptly revalued their currencies relative to others. This did not affect the value of money within a given country, so there is symmetry here: the value of home-made goods and services within a country are invariant under re-scaling of the exchange rate. In 1967 the British Prime Minister Harold Wilson devalued the pound against the US dollar, and was ridiculed for his statement that this adjustment would not affect 'the pound in your pocket'. But he was absolutely right in the sense that the relative values of goods, services and money within the UK were unaffected. The trouble came, of course, from the fact that imported goods went up in price, so people were generally worse off. A different sort of re-scaling occurs when inflation runs away, and prices have to be expressed in ridiculously large numbers. Governments will sometimes recalibrate their entire currency, as when the French introduced *le nouveau franc* in 1960. The new franc was worth 100 old francs, but the actual value of goods and services remained unchanged.

In physics there are many abstract symmetries. Positive and negative electricity provide a simple example. They stand in the same relation as left–right symmetry: if all the positive and negative charges in the universe were magically transposed, the laws of electromagnetism would be oblivious to the switch. Or to give another example, according to Newtonian mechanics the energy needed to lift a weight from the bottom of a building to the top depends on how tall the building is, but not on whether we choose to measure heights from sea level or ground level (the energy depends only on the height *difference*, not the absolute height). Nor would the energy be affected if we decided to switch our units of measurement from centimetres to metres, as with *le nouveau franc*. (In fact, the symmetry is even greater than I have indicated, because the energy expended in lifting a weight doesn't depend on the actual path taken either. It could be straight up or zigzag – it makes no difference to the answer.) The latter type of symmetry, in which a quantity is left unchanged when the measuring system is re-scaled, or re-gauged, is very common in physics, and is technically termed a *gauge symmetry*. Electric fields have a similar gauge symmetry: voltages can be re-gauged by adding or subtracting a fixed number of volts everywhere without affecting the energy change involved in transporting electric charge from one place to another; nor does the energy change depend on the path taken. So both Newton's laws and the laws of electricity are unchanged by such gauge transformations: this is a very basic property of these laws. Physicists have found that generalizations of such gauge symmetries go a long way to capturing the essential properties of the four forces of nature. Indeed, the forces are best classified precisely by specifying their symmetries, using a branch of mathematics known as group theory.

A classic application of the foregoing ideas is provided by the electroweak force. In Chapter 4, I explained how the Glashow–Salam–Weinberg (GSW) theory successfully unifies the electromagnetic and weak forces, and that at high energy the two forces blend together. Abstract gauge symmetries are interwoven into these forces, and it is the spontaneous breaking of one of these symmetries that leads to the differentiation of the two forces at low energy. In the context of the big bang, what happens is this. At temperatures above 10^{15} (a thousand trillion) degrees, the underlying unifying symmetry was manifested. The weak force had a long range, like the electromagnetic force. Then, as the temperature fell, there was a phase transition at which the weak force gauge symmetry was spontaneously broken, as a result of which the weak force abruptly became very short-ranged and considerably weaker than the electromagnetic force. And that is what we find in the relatively low-energy world of everyday physics. For many years physicists assumed that they were dealing with two separate forces because the symmetry that links them was broken. In fact, the physicist Enrico Fermi developed a theory of the weak force based on this misunderstanding, one which involved a very different law than appears in the GSW theory. But we now know that the broken-symmetry phase of the weak force does not reflect a truly fundamental law of physics at all, but a frozen accident, and Fermi's theory is just an effective theory, valid at low energy only.

The Higgs mechanism

You might be wondering how the electroweak symmetry gets broken. We saw in Chapter 4 that the exceptionally short range of the weak force can be explained by the exceptionally large mass of the W and Z particles, which are exchanged to convey this force; and, earlier in the present chapter, how subatomic particles are believed to acquire their masses by interacting with an all-pervading field – a so-called Higgs field. Well, it is the same Higgs field that is responsible for breaking the weak force's gauge symmetry. This is how it happens. At temperatures above that of the electroweak phase transition, such as occurred in the hot early universe before about a trillionth of a second, the Higgs field averages to zero, rather like the magnetization of the iron bar above the Curie temperature. With a zero average Higgs field, all the particle masses are zero. The W and Z particles are massless, just like the photon, implying that both the weak force and the electromagnetic force have long range. When the temperature drops, the Higgs field faces the same conundrum as the pencil standing on its head (see Figure 24, on p. 180): a trade-off between symmetry and stability. This is because the Higgs field interacts with itself, and the energy of this coupling resembles the gravitational energy of a poised pencil. A zero Higgs field is symmetric but unstable; a non-zero Higgs field is stable but breaks the symmetry. What does the Higgs field do? Like the toppling pencil, it spontaneously breaks the symmetry, as a result of which its strength jumps from a zero average value to a very large non-zero value. Some of the 'toppling energy' thus liberated gets used to pay for the masses that the Higgs field bestows on all the participating particles. It turns out, however, that in the conventional version of the Higgs theory, 'whichever way the pencil topples', the particles acquire the *same* mass. So although the electroweak phase transition in the early universe may have produced a domain structure of some sort, it isn't likely to be one in which particle masses vary from one domain to another. To get *that* sort of effect, it is necessary to consider higher energies still.

In Chapter 5, I described attempts to unify the electroweak force with the strong nuclear force, within some sort of grand unified theory, or GUT. These theories also involve symmetry-breaking mechanisms, Higgs fields and abstract symmetries, but they are more complicated and elaborate. Although we are now in much more speculative territory, the general principles I have outlined may still be applied. We can expect that at times earlier than a trillionth of a second and at correspondingly higher temperatures, there would be more symmetry, and less structure and complexity. For the record, the GUT temperature (at which any broken symmetries would be restored) is about a trillion trillion degrees, and corresponds to a cosmological epoch before about 10^{-36} s (a trillion-trillion-trillionth of a second).[10] As the universe cooled from this torrid condition, various symmetries would have broken in a series of phase transitions, probably forming huge cosmic domains. One of these transitions broke the symmetry between matter and antimatter. Another may have broken supersymmetry. Whatever the gory details, the bottom line is this: *both* the physical states (i.e. the nature and form of matter) *and* the effective low-energy laws would have become progressively more complex and differentiated as the universe cooled. From the point of view of the Goldilocks enigma, this complexity is a boon, because the rich domain structure predicted by GUT symmetry-breaking could have a major impact on the habitability of the universe. Domains might, for example, have different particle masses and force strengths, varying degrees of 'mixing' between different forces, and so on. But an even more exciting possibility now emerges . . .

The laws of physics might be just local by-laws

The way in which the more embracing GUT symmetries break down into strong and electroweak components might not be unique. There might be other ways of breaking the symmetries, leading not merely to different relative strengths of the forces, but to *different forces entirely* – forces with completely different properties than those with which we are familiar. For example, there could be a strong nuclear force with twelve gluons instead of eight, there could be two flavours of electric charge and two distinct sorts of photon, there could be additional forces above and beyond the familiar four. So the possibility arises of a domain structure in which the low-energy physics in each domain would be spectacularly different, not just in the 'constants' such as masses and force strengths, but in the very mathematical form of the laws themselves. The universe on a mega-scale would resemble a cosmic United States of America, with different shaped 'states' separated by sharp boundaries. What we have hitherto taken to be universal laws of physics, such as the laws of electromagnetism, would be more akin to local by-laws, or state laws, rather than national or federal laws. And of this pot-pourri of cosmic regions, very few indeed would be suitable for life.

Although GUTs offer some scope for a patchwork universe with drastically different laws in different domains, the possibilities pale into insignificance when compared with the richness of the string theory multiverse. There is now no shortage of possible low-energy domains: they emerge in their squillions. In fact, string theory opens up a veritable Pandora's box of possibilities. The stunning fecundity of this theory stems from the stupendous number of ways in which the extra dimensions can be compactified, or 'rolled up', in the manner I described in Chapter 5. Compactification is string theory's version of symmetry-breaking. For example, a simple symmetric shape such as a big, six-dimensional sphere might shrivel spontaneously into a complicated, multi-dimensional labyrinth of twisted bridges and bifurcating tunnels. One of these shapes, projected down onto a two-dimensional surface for ease of visualization, is illustrated in Figure 25. There are countless different configurations like this. The key point

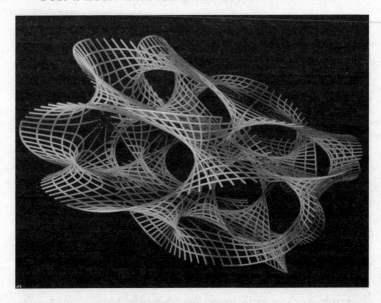

25. *The shape of dimensions unseen*

This complicated two-dimensional surface illustrates one among myriad 'compactifications' of the six unseen dimensions of space postulated by string theory. According to the theory, the specific shape determines the form of the physical laws in the remaining (large) three space dimensions. (Courtesy of Jean-Francois Colonna.)

here is that the laws of physics that apply in the remaining (uncompactified) space depend on the specific shape of the compactified dimensions. Let me spell this out, because it is so important:

> **In string/M theory, the low-energy physics of the three-dimensional world we observe is determined by the particular shape of the compactified extra dimensions.**

Which sort of universe we observe – such things as the nature of the forces (e.g. which symmetries they possess), the strengths and degree of mixing of the forces, the number of fundamental particles, their types (fermions, bosons), their properties (masses, electric charges, spins, etc.) and the energy of the vacuum (i.e. the dark energy) – all these things depend on the way in which the extra dimensions curl up. If they shrivel this way, you get a universe with five photons and two strong forces; fold that way, and there are eight electrons and four gluons. And so on.[11] In fact, even the number of compactified dimensions isn't fixed. There could be low-energy worlds in which only five dimensions undergo compactification, leaving a space of four large (visible) dimensions. Or perhaps seven get compactified, leaving only two large dimensions.

The string landscape

A challenge for string theorists is to determine how many different shapes the compactification process can create: how many possible low-energy worlds might be predicted by the theory. The answer is – a lot (maybe even an infinite number). In fact, even getting an estimate is hard because the mathematics of compactification isn't well understood. Added to this, one must take into account all sorts of further variables, such as how the loops of string are enmeshed in the topology, various abstract string symmetries and other refinements. Anyway, some estimates[12] put the grand total at more than 10^{500}. This number is one followed by 500 zeros! By comparison, the number of atoms in the observable universe is no more than 10^{80}. So there are incomparably more possible low-energy universes coming out of string theory than there are atoms in our universe.

A good way to envisage this mind-boggling cosmic cornucopia is in terms of a 'landscape' of low-energy universes, a helpful idea thought up by Leonard Susskind. (This is not a physically real landscape, but an abstract multi-dimensional 'landscape of possibilities.') Imagine a vast and complicated terrain of hills and valleys. Each point in the landscape represents a possible universe characterized by certain physical laws. The valleys in the landscape correspond to *stable* universes; ours would be one of them. Other valleys would correspond to different universes with different laws. Some might be not too different from ours, others would differ drastically. String theory yields so many different sorts of universe that the variations in their physical laws can be very fine indeed – almost continuous, in fact. There will be universes identical to ours except for the fifth decimal place of the electron's mass, or the tenth decimal place of the strength of the electromagnetic force. Others will have slightly larger differences in such quantities, while still others will be completely unlike ours, with new sorts of particles and very unfamiliar forces. It is not too much of an exaggeration to say that you could dream up a universe, choosing whatever sort of low-energy physics takes your fancy (within reason), and there will be a universe somewhere matching that description among the unimaginably vast smorgasbord of possibilities.[13]

Taken on its own, Susskind's landscape represents a vast selection of possible universes, but it does not predict that they actually exist. However, there is a natural mechanism for 'populating' the landscape with *really existing* universes. That mechanism is eternal inflation, described in Chapter 3. In this theory, an overall matrix of inflating space has no beginning or end; embedded within it are 'bubbles' which have ceased inflating and constitute pocket universes. One of them is ours. Pocket universes are forming all the time, 'nucleating' out of the eternally inflating space, rather like bubbles of dew nucleating around dust particles in water-saturated air. Each nucleation event represents the genesis of a separate pocket universe. Within a given pocket universe there will be a unique cosmic story: a hot big bang, cooling, symmetry-breaking, phase transitions and the emergence of a low-energy world of physical laws.

Combining eternal inflation with string/M theory, we can ask the following: if pocket universes pop out of the inflating matrix, where in the landscape of possibilities will the bubbles nucleate? Will they always appear at the same place, leading to the same low-energy physics, or will it be different every time? The theory suggests the latter. Because the nucleation events are quantum mechanical in nature, there will be unavoidable inbuilt variations. Furthermore, even within a given bubble, it is possible (though normally exceedingly rare) for a quantum fluctuation to nucleate another bubble at a lower energy, and another within that, and so on – bubbles within bubbles within bubbles, the outer ones expanding faster than the inner ones to give everything plenty of elbow room. Because inflation is eternal in this theory, there is an infinite time for the quantum mechanics to 'explore' the entire landscape in this manner.[14]

If these ideas are right, then the multiverse is populated by countless pocket universes in which all possible low-energy worlds – all 10^{500} of them – are represented somewhere. Cosmology is thereby transformed into an environmental science, in which a basic part of the explanation for what we observe in the universe depends on features of the local cosmic environment. In estate agents' jargon, it all boils down to location, location, location. (The theories I have described here are by no means the only ideas for a multiverse.[15])

Many scientists hate the multiverse idea

In spite of its widespread appeal, and its apparently neat solution of the Goldilocks enigma, the multiverse has some outspoken critics from both inside and outside the scientific community. There are philosophers who think that multiverse proponents have succumbed to fallacious reasoning in their use of probability theory.[16] There are many scientists who dismiss the multiverse as a speculation too far. But the most vociferous critics come from the ranks of string theorists, many of whom deny the validity of a landscape of vastly many worlds. They expect that future developments will expose this mind-boggling diversity as a mirage, and that when the theory is fully understood it will yield a unique description – a single world, *our* world. So far, however, there is little or no evidence to support that viewpoint; it remains an act of faith. Nevertheless, the backlash against the multiverse idea has been fierce. Prominent scientists and commentators have used words such as 'fantasy', 'virus' and 'intellectually bankrupt' in their denunciations. Paul Steinhardt, Albert Einstein Professor at Princeton University, finds the entire concept so distasteful that he has simply closed his mind to it: 'This is a dangerous idea that I am simply unwilling to contemplate,' he has declared.[17]

What lies behind such emotional outbursts? For those theoretical physicists hard at work trying to formulate a unique final theory, the multiverse comes across as a cheap way out. Transforming cosmology into a messy environmental science looks a shabby let-down when set alongside the inspiring magnificence of a unique final theory that would explain everything. Purists dream of finding deep reasons, bolstered by elegant mathematics, for why the world is the way it is – in all its myriad details. By contrast, the multiverse theory declares that the only reason we observe what we do is because it is observable. Randomness plus observer selection strikes many physicists as an ugly and impoverished explanation compared with an overarching mathematical theory that pins down the properties of the world with quantitative precision, and interweaves them into a harmonious unity. They regard multiverse/anthropic explanations as serving to undermine the efforts of the unification programme (e.g. string theory) and threatening its funding base. Some critics even go so far as to suggest that it perverts the education of young researchers.

A champion of the anti-multiverse faction is David Gross of the University of California at Santa Barbara, a Nobel prize-winning physicist who helped to develop quantum chromodynamics. Gross is an irrepressible optimist who believes that one day we will put together a final theory of everything that explains all the parameters of physics and cosmology in terms of well-understood mathematical laws. I have described the path to a theory of everything as a progressive unification of physics, a process in which seemingly different and independent laws are found to be linked at deeper conceptual levels. As more of physics falls within the compass of unification, there are fewer free parameters to fix, and less arbitrariness in the form of the laws. It isn't hard to imagine the logical extreme of this process: all of physics amalgamated into one streamlined set of equations. Maybe if we had such a theory, we would find that there were no free parameters left at all: I shall call this the 'no free parameters' theory. If that were the case, it would make no sense to consider a world in which, say, the strong force was stronger and the electron lighter, because the values of these quantities wouldn't be independently adjustable – they would be fixed by the theory. There are certainly some gung-ho string/M theorists who foresee a future development of the subject in which numbers such as 1,836, the ratio of the proton to the electron mass, and 10^{40}, the ratio of the electromagnetic to the gravitation force, will emerge from a welter of breathtaking mathematics. But at the present time this is just promissory triumphalism. String theorists are a long way from explaining even one of these numbers. In spite of this lack of progress, many of them remain sanguine. Confronted at an international conference with the challenge that a satisfactory final theory seems a distant prospect, Gross replied stoically, paraphrasing Winston Churchill: 'Never, never, never give up!'

Such fractious debate, however, does not imply that the multiverse theory turns on the validity of the string theory landscape. To be sure, the string landscape provides the richest and most natural version, but some form of multiverse is a generic feature of a hot big bang cosmic origin combined with symmetry-breaking. A universe which cools from an ultra-hot initial state will almost inevitably form a domain structure in which different domains have different properties, including low-energy effective laws and values for some of the constants of nature. Although the term 'multiverse' was coined relatively recently, speculations about a multiplicity of cosmic domains based on GUTs, higher-dimensional theories and other attempts at unification have been around for three decades. In the absence of a convincing unique final theory, the default assumption is that the universe we observe is merely one fragment among a haphazard patchwork of universes.

But is it science? Can the multiverse theory be tested?

Another frequently voiced criticism of the multiverse is that it isn't science because it can't be tested by experiment or observation. This objection has some force. The claim that our universe is accompanied by countless others looks impossible to check. I have already explained that in the theory of eternal inflation we cannot directly observe the other pocket universes, for two reasons: because they are unbelievably far away, and because they are receding from us much faster than light. It can be validly objected that a theory which rests on entities that are in principle unobservable cannot be described as scientific.

It is conceivable, however, that *indirect* evidence could be found to support the theory. Sometimes in science one can have confidence in a prediction made by a theory even when that specific prediction is untestable, if the theory as a whole enjoys good experimental support. For example, the general theory of relativity can be applied to the interior of black holes, regions of space we cannot observe from outside, even in principle, because they are surrounded by an event horizon. Yet general relativity has been so well tested in other contexts that physicists are fairly confident they can use the same theory to describe what happens inside black holes too. If string/M theory, or some other unified theory which predicts a multiplicity of possible worlds, could be sufficiently well tested experimentally, one might become confident about the theory's prediction of a multiverse. Unfortunately, an experimental test of string theory seems a distant prospect, but this is just a limitation of the current state of scientific endeavour. There is no fundamental reason why, in the far future, a complete unified theory could not be worked out and tested experimentally. So in this respect the multiverse theory hovers on the borderline between science and fantasy.

Indirect evidence of a multiverse could also come from investigating the details of fine tuning. The multiverse theory seeks to replace the appearance of design by the hand of chance. The advantage of chance is that it can be well defined mathematically. The rules of chance and the properties of random variables have been very thoroughly studied, and certain characteristic features are well known. The essence of the anthropic mechanism is that our universe is selected by us on account of its habitability, and in that respect it is highly atypical of universes as a whole. But such is the stupendous number of universes (at least in the string theory version of the multiverse) that there will still be a range of possibilities, including plenty of other universes which are fit for life but differ only marginally from ours. Fine tuning isn't an all-or-nothing affair: each of the relevant parameters enjoys a range of values consistent with life. A universe in which the electromagnetic force is, say, 1 per cent stronger would probably still be habitable, though 50 per cent stronger would cause us problems. Within the set of habitable universes, there is no reason to suppose that ours is anything but a typical member. So the multiverse theory makes the prediction that, when the fine tuning is researched more carefully, we should find that the measured values of the vital (i.e. bio-sensitive) variables display *typical* values within their life-permitting ranges.

How would we know whether the values were typical? A general feature of random processes is that big flukes are much rarer than little flukes. Think of tossing a coin ten thousand times. We expect the total number of heads and tails to be roughly equal after that many trials. However, it would not be surprising to have a run of three heads from time to time. Four heads would raise an eyebrow, five would be memorable and ten heads in a row, while not impossible, would be amazing. Such runs are called statistical fluctuations, and they are part and parcel of all random processes. The rule is, the bigger the fluctuation (in this case, the sequence of heads), the less likely it is to occur. Applying this rule to the multiverse, we find that there are likely to be many more universes (or cosmic domains) in which the circumstances for life are only just met than those in which the necessary conditions are met with a large margin to spare. In other words, there should be many more universes in which life is a close call than ones in which the degree of bio-friendliness is unnecessarily precise. Therefore, if we human beings are just arbitrary, random observers among all possible observers, then we are far more likely to find ourselves living in a universe only marginally bio-friendly than one which is optimally bio-friendly, simply because there are many more universes of the former sort than of the latter.

To take a concrete example, consider dark energy, the 'natural' value of which is 120 powers of ten greater than the observed value. As I have mentioned, Steven Weinberg suggested that this is an anthropic selection effect: our universe is a fluke as far as dark energy is concerned, and has been selected by us for its habitability (galaxies would not form if the dark energy were much larger). Universes with such strong suppression of dark energy are, in this theory, very rare. Applying the rule that small flukes are much more likely than big flukes, there should be many universes with dark energy values close to the enormous natural value, somewhat fewer with values modestly less than the natural value, and far fewer with dark energy very much less than the natural value. We might then expect our universe to lie close to the limit of what is a life-permitting value, on the basis that there are very many more universes of that sort than universes with even lower values of dark energy. And in fact this is not far off what is observed. The measured value of dark energy is probably no smaller than one-tenth of the 'lethal' value – the value above which galaxies would not form. If the observed dark energy were, say, a million times smaller than the maximum value allowed for life, the multiverse explanation would be decisively ruled out, on the ground that the universe would then be far more bio-friendly than it needs to be to explain our small-dark-energy universe as an anthropically selected statistical fluke.[18]

The reasoning used above might be easier to grasp with the help of an analogy. Imagine a lottery in which a large prize is given for anyone who correctly guesses four out of five randomly selected numbers between one and ten. The police warn the organizers to beware, because they have heard rumours of a conspiracy to cheat. The results are examined, and one man is found to have correctly guessed all five numbers. Nobody else has fulfilled the minimum requirement for a prize. The organizers are immediately suspicious, because the man need only have guessed four of the five numbers. So his entry was much better than it needed to be to win the prize, because guessing all five numbers is about ten times as hard as guessing only four. If the results had shown, say, eight people correctly guessing four numbers and one person with all five, the organizers would have regarded this as in accordance with expectations for random chance. But a single 'overkill' entry raises suspicions that something fishy is going on behind the scenes.

In the same way, if a physical parameter vital for life is ten times more bio-friendly than it needs to be for us to exist, this too should raise suspicions that random chance is not the explanation and that 'something fishy is going on' behind the cosmic scenes. A margin of ten – the current estimate for the case of dark energy – is a bit too big for comfort (i.e. to say that the measured value of dark energy is 'close' to the life-permitting limit when it is a factor ten smaller is a bit of a stretch). However, the theory of galaxy formation is complicated and still not fully understood, and it may be that with more research it will be found that dark energy just two or three times higher than the observed value would be enough to inhibit life. Anyway, the multiverse theory makes the prediction that galaxy formation (or some other process affecting life) should be frustrated by a value of dark energy only moderately larger than the measured value. If that turns out to be an incorrect prediction, then it would falsify the multiverse theory and point instead to 'something fishy'. And a theory that is potentially falsifiable is considered by most scientists to qualify for the description 'scientific'.

Have varying 'constants' already been detected?

There is a third way in which the multiverse theory might be testable. Although we may not be able to observe other universes with other laws, or with different values of the physical constants, we might be able to observe tiny variations of the laws *within our* universe. If such variations existed they would falsify the claim that the laws are uniquely fixed by a final theory. Clearly, if the laws can change slightly within the observed universe, they can change a lot in the regions beyond. Astronomers and physicists have performed very careful tests to look for any hint that the laws may have varied over time, or across space. They have been able to place very stringent limits on any such variations. For example, if Newton's gravitational constant G, which sets the strength of the gravitational force, were to change with the age of the universe, it would show up in how the planets orbit the sun. No such changes have been observed. Limits have also been placed on possible variations in the strength of the weak force and in the electron/proton mass ratio by studying a curious geological formation in West Africa. A deposit of uranium ore in Gabon is so rich that 2 billion years ago it went critical, forming a natural nuclear reactor. By examining the fission products, physicists have been able to confirm the 'constancy of the constants' involved in the reactions to very high accuracy.

Recently there has been a claim, albeit controversial, that the strength of the electromagnetic force has changed very slightly, by a few parts in a million, since about 6 billion years ago. The evidence comes from studying the fine structure of spectral lines from very distant quasars. If the force that binds electrons to nuclei changes, it affects the details of the 'barcode' that identifies atoms uniquely, and it is this that some astronomers think they have detected. Should the observations hold up, this would support the idea that at least one of the fundamental 'constants' of nature isn't absolutely fixed, but has the character of an environmental variable.[19]

On the scientific front then, the multiverse is not wholly speculative. It is rooted in respectable science and has the possibility of being indirectly tested. The general idea of a multiverse seems to be an inescapable feature of big bang cosmology combined with particle physics. But how far down this slippery slope should one slide? Once we are into the business of postulating entire other universes on flimsy evidence, where do we stop? And what does it do for our understanding of reality?

Universes galore: the problem of duplicate beings

A well-known aphorism says that in an infinite universe anything that can happen must happen. Subject to some mathematical pedantry, that is true. A simple example is coin tossing. The odds against flipping a coin and getting heads 1,000 times in a row are almost infinitesimally small (about one chance in 10^{301}). Nevertheless, if there are enough coins being tossed, it will happen somewhere. To get a feel for the numbers involved, think of every atom in the observable universe as a coin, being flipped once a second. The odds against a sequence of 1,000 heads showing up among the 10^{80} atoms during the age of the universe are still less than one in 10^{200}. The longest run of heads you might expect, even tossing at a trillion times a second, is about 360. If an infinite number of coins are being tossed, however, getting 1,000 heads in a row is absolutely certain: infinity will beat any odds, however adverse. In fact, 1,000 heads will crop up not just once, but an infinite number of times.

The same basic statistics apply to any system in which chance plays a role, such as the structure of the universe. Imagine that space is infinite and that the observable portion of the universe is typical of the whole (in contrast to the multiverse theory). If you travel far enough in any given direction you will eventually find a planet very similar to Earth, because the same sort of processes will be happening in the same sort of way. Keep on going, and after a colossal number of near-Earths you will eventually come across one that is almost identical to Earth, with the same continents and mountain ranges and oceans. It's straightforward to work out the odds of another Earth forming by chance, and hence the expected distance you would need to travel to find such a duplicate planet. The more nearly the other Earth resembles ours, the smaller the odds and the farther away it will be. It may seem bizarre to imagine another planet just like ours, but in a truly infinite universe other Earths will be out there for sure, albeit a fantastic distance away. By extension, there will be other Earths with other beings like humans. And lengthening the odds still more, we can conclude that there will be other Earths, not just with other humans, but with other *yous*, identical in all respects, including your lifetime experiences. But by the remorseless logic of statistics, for every Earth containing an identical you there will be countless more Earths with a being differing from you only in some slight respect, such as hair colour or height, or last year's birthday present.

The cosmologist Max Tegmark of the Massachusetts Institute of Technology has worked out that the average distance to the nearest identical you should be about 10 to the power 10^{29} metres, under the assumptions made (e.g. uniformity of the laws of physics and the distribution of galaxies). Compare that with the diameter of the observable universe, a paltry 10 to the power 26 metres. Clearly, the risk that you are ever going to run into one of your alter egos is negligible. In fact, in a standard big bang model universe the separation between identical beings is typically far, far greater than the horizon size, meaning there can be no possible contact or even exchange of messages between them during their lifetimes. But even if you need not fear an encounter with a duplicate you, the very notion that there could be not just one, but an *infinity* of identical copies of you, leading identical lives (and infinitely many others leading similar but not identical lives) is deeply unsettling. Even Tegmark admits that his gut reaction is to find the idea 'strange and implausible'.[20] The problem concerns personal identity. We feel unique. If there is another me, let alone countless other mes, would they be *me* or somebody else? Such questions make our heads spin.

It is but a simple extrapolation of these ideas to work out how far you would have to go before running into an entire observable universe identical to our own. The answer is about 10 to the power 10^{120}

metres. But by the same reasoning, these carbon-copy universes *must* be out there somewhere, if the universe is truly infinite. And there will be infinitely many of them: infinitely many universes identical *in all respects* to the observed universe. Weird though these conclusions may seem, they follow incontrovertibly from the logic of simple statistics and probability theory.[21]

Pivotal in this analysis is the assumption of an infinite universe that does not differ systematically from one region of space to another. How reasonable is this? The popular theory of eternal inflation suggests something quite different: that our universe is not infinite; rather, it is embedded in a huge but nevertheless finite region similar to our own. Vast though the inflation region (or pocket universe) may be, it is nowhere near vast enough to accommodate more than one you with any significant likelihood. But this is a pyrrhic victory, because the finiteness of our inflation region is bought at a price, that price being the existence of an infinite number of *other* cosmic regions, or pocket universes. So all we are doing is trading one infinity for another. Eternal inflation predicts that there will be an infinity of universes produced by the universe-generating mechanism. There will be big ones and little ones and, if the string landscape ideas are right, universes with all possible variations of laws and initial conditions. There will also be infinitely many universes resembling ours, and, buried among this infinity, there will be infinitely many universes *identical in all respects* to ours.

Is the universe a fake?

As if the prospect of infinitely many identical universes doesn't already play havoc with our concept of reality, worse lies in store. One of the biggest box office successes of the last few years was *The Matrix* series of movies, and since much of the filming was done in Sydney, where I used to live, I feel a certain affinity. The theme running through this sci-fi drama is that (roughly speaking) human beings may not be real, but the product of a computer simulation. What the characters in the film take to be the real world is actually a virtual reality show, created by an advanced civilization with enormous computing resources.

The idea that the world about us is some sort of fake, or a simulation designed to fool us, has a long history in philosophy and science fiction, where it is used as a thought experiment to instruct, entertain or perhaps bemuse. What is new is that some eminent scientists and philosophers are now asking us to take the simulation idea seriously. To put it bluntly, they are conjecturing that the universe and the observers it contains – and that includes *you* – may be the product of a gigantic computer simulation being run by an unknown being or beings. Nick Bostrom, an Oxford philosopher and pundit of the so-called *simulation argument*, spells it out: 'There is a significant probability that you are living in [a] computer simulation. I mean this literally: if the simulation hypothesis is true, you exist in a virtual reality simulated in a computer built by some advanced civilisation. Your brain, too, is merely a part of that simulation.'[22]

Wow! What Bostrom is saying is that the objects in the room about you, the chair on which you are sitting, this book you are reading, and the people you know and love, the solid matter of your body – not to mention the sun, the stars and the rest of the universe – are all a figment of your imagination. Worse still: *you* are a figment of your imagination. And this mental experience – this 'imagining' – is concocted inside some vast computer which exists not in this universe (which doesn't 'really' exist, it is only virtual reality), but in some hypothetical 'other' universe, or 'master' universe. Is this even remotely credible?[23] Well, once we are prepared to entertain the possibility of a multiverse, there seems to be no good reason to rule out universes which contain computer simulations of other universes. In that case, the multiverse will be populated by both real and simulated universes, and any serious analysis of the pros and cons of the multiverse theory cannot escape the bizarre consequences of the existence of fake universes.

The closest most of us come to the *Matrix* experience is dreaming. Some dreams are so vivid that they seem at the time to be real. I have even had dreams within dreams, where I have been sure that I have woken from a nightmare only to discover, later, that I was still asleep. Some people have dreams (of extraordinary experiences such as alien abductions) which are so convincing that they remember them afterwards as factual experiences and report them as such, even under close questioning and hypnosis. Mostly, however, we can distinguish the dream world from the real world, and we are clear about which is which: the dream world is not real – it is a (usually rather poor) simulation, or illusion, created in our restless brains by some process that is still not well understood. But it's not difficult to imagine a dream world so coherent and vivid that it approaches 'the real thing', leaving us hard pressed to say whether we were dreaming or not. Gottfried Leibniz was ahead of his time in offering such a speculation: 'Although the whole of life were said to be nothing but a dream and the physical world nothing but a phantasm, I should call this dream or phantasm real enough if, using reason well, we were never deceived by it.'[24]

The goal of the virtual reality industry is to create sensory illusions so realistic that the subject perceives the virtual world as if it were real. This is accomplished by using devices such as three-dimensional visual displays, immersive sound, suits which deliver tactile stimuli, and gloves which move images through electronic signalling as if the wearer were touching a real object. Virtual reality done this way is not quite the same as dreaming, which is confined to the inside of our heads. But a future technology might be capable of delivering electronic signals directly to a subject's brain, creating the sensory impressions of a fake external world without the bother of stimulating the eyes, ears, and other senses. A favourite science fiction theme revolves around a disembodied brain incarcerated in a vat, wired up to some fancy computer system which creates the impression of a real world, so that the subject whose brain it is might be completely unaware that they no longer have a body and that their perceived world is a fake. Indeed, this idea is so familiar to philosophers, who like to use it to discuss the nature of observation and reality, that the term 'envatment' has entered their lexicon.[25]

The ultimate simulation is one that does not use existing brains at all (in contrast to dreams and envatment) but simulates the consciousness of inhabitants of the virtual worlds directly. To buy this idea, one has to accept that consciousness is not some sort of essence, but is the product of physical processes. This is certainly the orthodox view among scientists. For a long time the more adventurous members of the so-called artificial intelligence (or AI) community have been telling us that one day computers will be powerful enough not only to calculate payrolls and play chess, but actually to think and be aware of their environment and their own existence. In short, they will acquire consciousness. This line of reasoning goes back to Alan Turing, the British mathematical genius and co-inventor of the electronic computer, who in 1952 wrote a famous paper addressing the question 'Can machines think?' Turing set out the criteria by which one would be able to answer 'Yes'.[26] And again, science fiction has been quick to depict sentient robots and androids with an 'inner life' like humans.[27]

Some scientists, most notably the Oxford mathematician Roger Penrose, have argued that the machines we currently call computers can never, even in principle, simulate consciousness.[28] I won't go into the details of this argument, or the many attempted refutations. The point is that even Penrose does not deny the possibility of simulating consciousness with *some* manufactured physical system, it's just that, according to him, the system would not operate as a conventional digital computer. Obviously in principle one could build an artificial brain by copying a natural one in detail and assembling it molecule by molecule. If one did, and embedded it into a body, it's difficult to imagine any overriding argument as to why this manufactured brain would not support true consciousness.

Would you know if you were living in a simulated universe?

If we accept that consciousness can be simulated, at least in principle, it is then only a small step to imagining that something like a conscious human being could be simulated, and indeed an entire community of such sentient beings, all of them existing, so to speak, *in silico*, or at least in some sort of complex manufactured machine or system. And this is precisely what many scientists and engineers believe humanity itself may achieve within, if not decades, then at least centuries (Moore's law permitting[29]). The obvious question then arises of how we can be sure that we ourselves are not the product of such a simulation. How can we tell whether we are real, or just 1's and 0's inside a super-civilization's supercomputer?

The short answer is that we can't – at least not from a casual inspection. The computer world *transcends* the virtual world it simulates. If we, and the world we perceive, are the creation of an elaborate machine processing vast amounts of information, then we are no more capable of apprehending this transcendent simulating system than a computer program is capable of knowing the computer it is running on. Software and hardware belong to different universes, or at least to different conceptual levels. That is why we don't 'feel' our brains – we are not aware of them – even though we 'live inside them'. By the same token, our consciousness might live inside a purely simulated brain living inside a superdupercomputer.

Even if we can't be sure whether the world about us is real or faked, we can at least contemplate the relative probabilities. How likely is it that the universe is a fake? The key point here is that fake universes are incomparably cheaper than real ones. To make a fake universe you just have to process bits of information, and although that costs some energy (computers get hot), it is far less than the energy needed to make 10^{50} tonnes of matter. Moreover, it's not necessary to make a whole universe to convince you and me that the world about us is real. Most of the universe goes unseen in daily life, so it can be dispensed with: just Earth and its immediate environs would suffice. Nor does the fake universe have to be billions of years old: the simulation could start at any time with records and memories intact. The simulation need not be precise in every detail. So long as we didn't notice the scenery wobbling, we would still be unaware that we were living in something like *The Matrix*. Given these facts, it is clear that one super-civilization inhabiting a real universe could at relatively little cost simulate an almost limitless number of fake universes. In other words, the ratio of fake universes to real ones is likely to be enormous.[30] If, then, a given sentient being cannot distinguish a simulation from reality, the vast majority of such beings are likely to be living in simulations. It follows that you and I are almost certainly simulated beings living in *The Matrix*. This is broadly Bostrom's conclusion.[31]

The multiverse must include fake as well as real universes

These wacky ideas might have remained restricted to the confines of esoteric philosophy classes if the multiverse theory hadn't risen to prominence. By its very nature, the multiverse challenges our understanding of what is real and what is imaginary. If we are prepared to entertain the notion that there exist limitless universes that are unobservable from this one, why should we rule out the existence of limitless simulated, or fake, universes too? No reason at all. In fact, not only have we no reason to rule them out, we have every reason to rule them in. While the simulation argument was restricted to a single universe, it was always possible to wriggle out of the uncomfortable conclusion that *this* might be a simulation by arguing that no civilizations are likely to reach the point of achieving such stupendous computational power. For example, there are many reasons why humanity may not survive for more than a few centuries beyond the present, and that may not be long enough for conscious computers to be developed.[32] If a similar fate were to befall any other intelligent beings who might be located elsewhere in the universe, then simulations, while still a possibility in principle, might never be achieved in practice.

But if our universe is part of a multiverse, the balance of probability shifts dramatically in favour of simulation. It's a matter of basic statistics. It may be true that humanity will never survive to the point of creating a super-civilization with immense computational power, and it may also be the case that no other planet in our observable universe achieves this level of development either. But the multiverse allows all possible variations on a theme, including similar but slightly different universes in which humanity does survive to the point of being able to simulate fake realities. Unless there is some law that forbids the emergence of such civilizations, not just in this universe but in them all (and it's hard to think what sort of law that might be), it is inevitable that some universes like ours will give rise to universe-simulating super-civilizations. These universes will then spawn a vast number of fakes, so that in the total mix of real and fake universes, fake ones will overwhelmingly predominate. Therefore our universe is very, very likely to be a fake. To be sure, creating fake universes might be regarded as a waste of time and resources by many civilizations, but it requires only *some* civilizations – for reasons of research, entertainment or altruism – to do it, and the fake worlds would proliferate.

Although we may not be able to tell from superficial appearances whether or not the observed universe is real or faked, a more careful inspection might disclose its artificial nature. The cosmologist John Barrow has speculated on this topic. 'Even if the simulators were scrupulous about simulating the laws of Nature, there would be limits to what they could do,' he writes:

Assuming the simulators, or at least the early generations of them, have a very advanced knowledge of the laws of Nature, it's likely they would still have incomplete knowledge of them . . . These lacunae do not prevent simulations being created and running smoothly for long periods of time. But gradually the little flaws will begin to build up. Eventually, their effects would snowball and these realities would cease to compute. The only escape is if their creators intervene to patch up the problems one by one as they arise.[33]

What, then, might give the game away if the simulators attempt to fix up their botched universe? Barrow points to the possibility of 'glitches' in the laws of physics:

In this kind of situation, logical contradictions will inevitably arise and the laws in the simulations will appear to break down now and again. The inhabitants of the simulation – especially the simulated scientists – will occasionally be puzzled by the experimental results they obtain. The simulated astronomers might, for instance, make observations that show that their so-called constants of Nature are very slowly changing.[34]

I mentioned in the previous section how there may indeed be some observational evidence for a slight tweak in the strength of the electromagnetic force about 6 billion years ago. Was this the quality controller of the Cosmic Simulation System literally fine-tuning our universe, resetting the dial on the Designer Machine? Barrow concludes 'that if we live in a simulated reality we should expect occasional sudden glitches, small drifts in the supposed constants and laws of Nature over time, and a dawning realisation that the flaws of Nature are as important as the laws of Nature for our understanding of true reality.' The implication being, of course, that the astronomical evidence from quasar spectral lines might already have spilt the beans.

The conclusion – if it can be taken seriously – that we are probably living in a computer simulation has some intriguing, not to say disturbing ramifications. If we exist care of an unknown and unknowable information processing system, what guarantee is there that the simulators in charge of it will continue to run the simulation? Perhaps they will get bored and switch it off?[35] Or run out of funding for their 'experiment'? And because the nature of the simulated world we experience is entirely up to 'them', what is to stop them from changing the software and transforming our elegant universe into an ugly or a hellish one?

It gets worse. Simulation lies at the heart of the theory of computation. The very notion of a universal computer (often called a Turing machine) is that it can simulate any other universal computer: this is known as the principle of computational universality. To bring it down to mundane practicalities, a Mac can simulate a PC, and vice versa. For this reason, a simulated world rich enough to include general-purpose computation (and big enough to provide the resources) can itself simulate its own virtual reality. This would be like the dream-within-a-dream I discussed earlier – which raises the scary prospect that we and our world may be a simulation being run by another simulation, two steps removed from reality! Logically there is no end to this nested sequence of simulations within simulations within simulations . . . The real universe could be lost amid an infinite regress of nested fakes. Or it may not even exist at all. Reality might consist of an infinite sequence of simulations, period.

Theology in a fake universe: there could be fake gods too!

The simulation argument has an amusing corollary. The status of the simulated beings is that their existence, and their world, depend entirely on the simulating system. If the superdupercomputer shuts down, or gets commandeered for another (less ambitious?) project, it would spell cyber-Armageddon for us all. So the simulating system is our transcendent designer, creator, sustainer and potential destroyer. This system could, if desired, communicate directly with us by using a conspicuous sign in our world, such as something being out of place or violating the normal laws, which would appear to us to be a miracle. Conversely, because the simulating system simulates our brains, it presumably has access to our thoughts, and so we can communicate mentally with it and with its operators – the simulators (who could of course be one and the same). The simulated beings thus stand in the same relation to the simulating system as human beings stand in relation to the God (or gods) of traditional religion. God is also transcendent designer, creator, sustainer and potential destroyer, plus miracle-worker, thought-knower and receiver of prayers. Since the multiverse argument is often invoked as a way to abolish the need for divine providence, it is ironic that it provides the best scientific argument yet for the existence of a god! Clearly, if a multiverse exists it is impossible to avoid the conclusion that at least some universes containing observers are the product of designer-creator gods. John Barrow has expressed it graphically:

gods reappear in unlimited numbers in the guise of the simulators who have power of life and death over the simulated realities that they bring into being. The simulators determine the laws, and can change the laws, that govern their worlds. They can engineer anthropic fine-tunings. They can pull the plug on the simulation at any moment, intervene or distance themselves from their simulation; watch as the simulated creatures argue about whether there is a god who controls or intervenes; work miracles or impose their ethical principles upon the simulated reality.[36]

The theology of this pantheon is bewildering. Not all universes will have gods. Sometimes gods will make uninhabited universes, and sometimes inhabited universes will be real rather than virtual, so will not have gods (at least, not transcendent ones). Universes can be the product of one god, or of many collaborating gods, or even of many competing gods (as in traditional polytheism). If the nested simulation argument is sound, then some gods will exist only by virtue of other gods having created them. There could be a hierarchy of gods – indeed, an unending sequence of gods – each one dependent for its existence on the god above. The gods I am describing here would be unlikely to fulfil the role of the God of monotheism, who according to classical Christian theology transcends and supports *all* possible realities. And of course, the existence of the multiverse as a whole, with its plethora of real and fake universes, is still left unexplained by the gods I have described. So the ultimate source of reality still eludes us.

Entertaining though these deliberations may be, there is a problem with the consistency of the entire argument. As I pointed out, there is no reason for the simulators to create a virtual universe which faithfully complies in every detail with the same laws that hold in the real universe. In fact, there is every reason not to bother, because very little of everyday human experience depends in any direct way on, for example, nuclear reactions or black holes. After all, my dreams are still half-meaningful without my brain being troubled about getting the laws of physics right. But the argument for the existence of a multiverse is based on the *perceived* laws of physics, laws we have discovered through scientific enquiry in this world. If the perceived universe is a fake, then so are its laws, and we have no justification at all in extrapolating the fake physics to the whole of reality; most especially we cannot assume that the simulators would bother to make countless unseen and unseeable universes alongside this one just to keep the fake physics in ours consistent. And since we would have no idea at all what the laws of physics might be in the *real* universe – and no reason to expect them to resemble 'our' laws – then we cannot assume that the real laws will permit a multiverse.

We are therefore faced with three alternatives. The first is to take it as given that there exists a real multiverse, rich and complex enough to give rise to universes like ours containing sentient beings. In which case the real multiverse almost certainly spawns a much more extensive fake multiverse, and we are very probably living in one of its fake components. The second option is to take it as given that we are living in a simulated universe, and accept that the fake physics we discover cannot be applied to problems of ultimate reality, and certainly cannot be used to argue for the existence of a multiverse, real or simulated. The third alternative is that there is just one real universe, or a restricted form of multiverse, and that some unknown factor prevents the simulation of consciousness.

So which is it to be? I first became embroiled in the subject of fake universes in a rather whimsical way. A few years ago I was invited to debate the existence of God with the Oxford University chemist Peter Atkins, a vociferous atheist. Such debates are so common they inevitably go over the same ground again and again. In the taxi to the lecture hall I was anxious to come up with something new and entertaining for the audience. I guessed that Atkins would invoke the multiverse argument as an explanation for why the universe appears to be designed for habitation, and I wanted to find an argument against this in the ten minutes it would take the taxi to arrive. It then struck me in a flash that an uncritical appeal to a multiverse opens a Pandora's box, by permitting fake universes to swamp the real ones. And since fake universes contain fake physics, the pat argument leading from the laws of physics to a multiverse to anthropic selection to the elimination of God would be neutralized in a self-contradictory loop. The multiverse advocates would be hoist by their own petard! Atkins handled this new challenge with aplomb, but I wasn't finished. I repeated the argument in an article in the *New York Times*,[37] pointing out that the threat of fake universes constitutes a *reductio ad absurdum* of the entire multiverse theory. The response from multiverse supporters was a big surprise. Rather than recoiling from the fake universe threat, they embraced the possibility with alacrity as part of an enlarged concept of the multiverse. Martin Rees expressed this new mood succinctly:

All these multiverse ideas lead to a remarkable synthesis between cosmology and physics . . . But they also lead to the extraordinary consequence that we may not be the deepest reality, we may be a simulation. The possibility that we are creations of some supreme, or super-being, blurs the boundary between physics and idealist philosophy, between the natural and the supernatural, and between the relation of mind and multiverse and the possibility that we're in the matrix rather than the physics itself.[38]

I remain cautious and sceptical about universes multiplying willy-nilly. Once the possibility of fake universes is accepted, there are strong arguments for concluding that ours is a fake, simply on the ground that the multiverse is likely to contain vastly more fake universes than real ones. While it may be true that our universe *is* a fake, it seems to me that drawing that conclusion would spell the end of scientific inquiry. In that respect it is akin to the argument that the universe was created five minutes ago with all records and memories imprinted in it – that the present is real, but the past is fake. We could not disprove this claim, but accepting it gets us nowhere. Belief that the universe is simulated resembles solipsism, the contention that only oneself exists. Solipsism is also impossible to disprove, because to obtain direct evidence that another mind exists you would have to become that mind – and then that person would be you. It is hard to argue with committed solipsists, for they simply assume that you are part of the grand conspiracy that mimics the existence of other minds, when in fact only one mind exists. Solipsists may be content in their conviction, but it makes no sense for a solipsist to try to convince anyone else. Similarly, belief that the universe is a conspiracy, faked to look like the real thing, gets you nowhere. There is no point in arguing such a case on scientific or logical grounds, because there is no reason why a simulated world should conform to scientific or logical principles – any more than cartoon characters should obey the laws of physics or the rules of logic. And in fact, once you accept that you yourself are a computer simulation, there is no good reason to suppose that the same simulation manufactures other minds, for how would your simulated world differ if you were the only sentient being and all the other beings in your fake world were just part of the simulation? So belief that the universe is a simulation is more or less equivalent to solipsism.

If one rejects the conclusion that the universe is a simulation, then does the *threat* of fake universes still constitute an argument against the multiverse (which was my original intention)? Here I am less certain. The simulation argument is predicated on the assumption that consciousness may be simulated by an information processing system, but that is an act of faith. It is not impossible, although it is unfashionable to suggest it, that consciousness is a fundamental quality which simply cannot be generated by a computer or suchlike. Even if consciousness can be digitally simulated, the simulation of universes containing observers may use up vast resources, negating the assumption that fake universes greatly outnumber real ones. If fakes constitute only a small fraction of the universes in a multiverse, then the argument that this universe might very well be a fake would be less persuasive. It therefore remains an open question whether fake universes are possible, and if so whether they would dominate the multiverse.

So far I have concentrated on two possible explanations of why the universe is so unexpectedly suited to life. The first is that any apparent fine tuning is simply an exceedingly lucky fluke, and there is no more to be said. The second is that there exists a multiverse, and the bio-friendly nature of the universe we observe is a selection effect. Both these explanations have strengths and weaknesses, and I shall return to them for a final assessment in Chapter 10. But first I need to deal with the third possible explanation of why the universe is bio-friendly, the explanation favoured by many non-scientists, and that is that the universe is so stunningly suited to life because it has been designed that way by an intelligent creator.

Key points

- Our universe may be a fragment of a vast (probably infinite) and heterogeneous system called the multiverse. The other 'universes', or cosmic regions, may be observationally inaccessible to us. Their existence would be inferred from theory plus some indirect evidence.
- The laws of physics and the initial state of the universe could vary from one 'universe' to another. What we have taken to be absolute laws might be more akin to local by-laws, with key features,

including those relevant to life, which 'froze' out of the hot big bang in the first split second.

- Physics as we know it is relatively 'low energy' compared with the heat of the big bang. As a general principle, cooling a physical system results in the breaking of symmetries and the emergence of complexity.

- The universe began simple, with simpler underlying laws. The observed masses of particles, for example, may have been acquired only in the cooler phase. Some features of the more complicated laws we experience now – features that resulted from symmetry-breaking – might be random. Therefore they could be different in other regions.

- Only where random chance produced the right laws and initial state will life and observers have emerged. Some scientists think that this might explain the uncanny bio-friendliness of the universe.

- Critics of the multiverse are very outspoken and damning. They include scientists who are striving for a final 'theory of everything' that would explain the universe completely without invoking a multiverse or observer selection effects. They hope for a unique 'one-world, this-world' solution to a unified theory.

- Some philosophers argue that simulated universes (e.g. virtual reality run on gigantic computers) may be possible. Multiverses might then include simulated as well as real universes. Some simplistic calculations hint that the fakes may greatly outnumber the real ones, so we could be living in a simulation!

9

Intelligent and Not So Intelligent Design

The watchmaker argument

Everyone agrees that the universe *looks* as if it was designed for life. Well perhaps it *is* designed for life. Perhaps there is a Designer? This is hardly groundbreaking reasoning – the design argument for the existence of God goes back hundreds of years. Augustine expressed the basic idea clearly when he wrote that, 'The very order, disposition, beauty, change and motion of the world and of all visible things silently proclaim that it could only have been made by God.'[1] In the thirteenth century, Thomas Aquinas chose evidence from design as his 'Fifth Way'.[2] The argument was popularized in the eighteenth century by an English clergyman named William Paley, who famously used an analogy between a watch and the wonders of the natural world.

Paley invited us to imagine that by chance we came upon a watch lying on the ground. Even without knowing precisely what it was, he suggested, we would soon deduce that it was an artefact designed for a purpose:

[W]hen we come to inspect the watch, we perceive . . . that its several parts are framed and put together for a purpose, e.g. that they are so formed and adjusted as to produce motion, and that motion so regulated as to point out the hour of the day; that if the different parts had been differently shaped from what they are, or placed after any other manner or in any other order than that in which they are placed, either no motion at all would have been carried on in the machine, or none which would have answered the use that is now served by it . . . the inference we think is inevitable, that the watch must have had a maker – that there must have existed, at some time and at some place or other, an artificer or artificers who formed it for the purpose which we find it actually to answer, who comprehended its construction and designed its use.

Paley went on to discuss 'the contrivances of nature' that are vastly more complex and design-like than watches: 'The marks of design are too strong to be got over. Design must have had a designer. That designer must have been a person. That person is GOD.'[3] In presenting evidence for a designer-God, Paley considered two types of natural system: astronomical and biological. The astronomical comes closest to the theme of this book, but the biological is better known and of some current interest, so I shall deal with that first.

Biological organisms are immensely complex – far more complex than Paley could have realized. To a physicist they look nothing short of miraculous. The many and diverse components function together in a coherent and amazingly orchestrated manner. The living cell contains minuscule pumps, levers, motors, rotors, turbines, propellers, scissors and many other instruments familiar from a human work-shop, all of them exquisite examples of nanotechnology. The entire assemblage runs itself with great efficiency, sometimes autonomously, sometimes in collaboration with other cells through a sophisticated network of intercellular communication based on chemical signalling. The command and control functions of the cell are encoded in its DNA database, which implements instructions through intermediary molecules using an optimal mathematical code to convert software instructions into hardware products with customized functionality. And this is just one cell. In a larger organism, vastly many cells get together and cooperate to form organs such as eyes, ears, brains, livers and kidneys, many of them immensely elaborate in their structure and function. The human brain alone has more cells than there are stars in the Milky Way galaxy. So it all adds up to a package of marvels that boggles the mind.

The appearance of design is one of the defining hallmarks of life. The question before us is whether living organisms actually *were* designed, or whether natural processes can mimic design well enough to explain what biologists observe. Darwin's theory of evolution, published in 1859, owes its success precisely to its ability to account for the appearance of design without invoking a designer (the so-called blind watchmaker argument, lucidly popularized by Richard Dawkins[4]). The theory is simple, and has been so thoroughly discussed in the literature that it needs only the briefest summary here. So: Organisms produce offspring with slight variations – taller, shorter, darker, lighter, slower, faster . . . Sometimes circumstances are such that one of these features is favoured (e.g. better to be faster if the name of the game is to escape predators), and organisms possessing that quality will have a greater chance of survival and of passing on their favoured genes to the next generation. As Dawkins graphically expresses it, good genes end up inside descendants, bad genes end up in the stomachs of predators. So nature acts as a sieve, filtering out maladapted genes and rewarding good genes with duplicates. In this way, incrementally, favourable traits become amplified and unfavourable ones eventually eliminated. Amplification of different traits in different circumstances leads to diversity, and when the divergence between similar organisms exceeds the threshold beyond which interbreeding can happen, they are considered to be separate species.

The only assumption made in framing this theory is that there will be variation, inheritance and selection. Selection operating in the struggle for survival is obvious to us all, but these days scientists understand how inheritance and genetic variation come about too, in terms of the molecular basis of life. Notice that although variations may be random, selection is far from random, so that it is not true to say, as is sometimes quipped, that Darwinism attributes the organized complexity of the biosphere to nothing more than random chance. It is obvious that chance alone would be no more likely to produce a living cell than (to use Fred Hoyle's famous analogy) a whirlwind blowing through a junkyard would produce a Boeing 747. But chance was *not* alone in fashioning the biosphere.

There is actually a fourth essential ingredient in Darwin's theory, and that is time. Selection can work only generation by generation, so changes tend to be slow and accumulate over immense durations. Billions of years are needed for life to evolve from a handful of simple microbes to the diversity of the biosphere we see today. But that's all right: Earth is over 4.5 billion years old. The fossil record, patchy though it inevitably is, provides overwhelming support for the fact that life has indeed evolved over at least 3.5 billion years from its humble origins in the form of simple microbes.

God-of-the-gaps makes a comeback

After some early skirmishes, most theologians came to accept Darwin's theory of evolution. They contented themselves with the belief that God could achieve his purposes by working (albeit slowly) *through* the evolutionary mechanism rather than against it. They conceded that although Paley's argument was right, his conclusion was flawed: God did not design and create the different species of living organisms one by one from scratch. Rather, they evolved gradually and incrementally as a result of variation and selection. Nevertheless, some critics of Darwin's theory pointed to certain organs, or organisms, which they considered to be so complex and so well organized that it seemed incredible that their existence could be accounted for by mere variation and selection.

A favourite example was the human eye, an organ which Darwin himself found baffling. The eye was held up as an illustration of *irreducible complexity*. The point here is not just that the human eye is a complicated organ, but that it contains many interconnecting and cooperating components, for example a lens, a light-sensitive surface and a pupil to control the influx of light. Remove just one of these components and the eye would be severely compromised. The puzzle seemed to be how the several parts, each of which individually is of limited use, could nevertheless come to be assembled in such an efficiently collaborative way. Since it is the essence of Darwinian evolution that selection acts gradually and incrementally to fashion new organs, and that each intermediate stage must have some selective advantage *at the time*, the eye seemed to provide a good example of a gap in the Darwinian account. Many other apparent gaps were identified.

The fossil record in Darwin's day was very patchy, which encouraged the belief in some quarters that God still had a role to play, presumably by popping up from time to time in evolutionary history, like a conjurer, to fix an unsatisfactory job: rearranging a few atoms here, tweaking a gene there. Many theologians, however, were not comfortable with this idea, as is well captured in the following remark by Henry Drummond, made over a century ago:

Those who yield to the temptation to reserve a point here and there for special divine interposition are apt to forget that this virtually excludes God from the rest of the process. If God appears periodically, He disappears periodically. If He comes upon the scene at special crises, He is absent from the scene in the intervals. Whether is all-God or occasional-God the nobler theory? Positively, the idea of an immanent God, which is the God of Evolution, is infinitely grander than the occasional Wonder-worker, who is the God of an old theology.[5]

So strongly did some theologians object to the idea of God as part-time biological tinkerer that they invented the derisory term 'God-of-the-gaps' to describe it.[6] The main objection to a God-of-the-gaps is not so much the happy-go-lucky – indeed, less than competent – nature of this style of designer, it was the ever-present risk that scientific advances would systematically close the gaps, squeezing God into smaller and smaller interstices, perhaps to be displaced altogether in due course. A God who lurks in the dark corners of human ignorance is a God who must make a slow and inexorable retreat as science progresses.

And many gaps have indeed been plugged. One of these is in fact the eye, the favourite example of irreducible complexity from the nineteenth century. It is easy to say that half an eye is useless, but that simply isn't true, as partially sighted people will testify. Any sort of light sensitivity is better than none, and on that basis it has been possible to reconstruct a plausible evolutionary history for the eye, starting with nothing more complicated than a light-sensitive patch, and requiring that each incremental adaptation offers some selective advantage over the previous set-up.[7] In fact, eyes have evolved many times, using many different 'designs', which suggests that it's actually not that hard for random variation and selection to do the job step by step, ratchet-like, accumulating many small changes. The Darwinian explanation is further bolstered by the fact that many of the intermediate stages in the development of complex eyes still exist in the animal kingdom: there are creatures out there using them!

Intelligent design in biology is magic, not science

In spite of the gap-closing that has been continuing for the past century and a half, supporters of a God-of-the-gaps simply move the goalposts and seek other gaps. A current favourite with the so-called Intelligent Design movement in the United States is the bacterial flagellum, an ingenious-looking device that propels the cell by a rotary action using a little motor. This system is claimed to have irreducible complexity. Quite why the bacterial flagellum is more 'irreducibly complex' than the eye isn't clear, and one would have thought that the lesson had been learned by now. Although a blow-by-blow account of the way in which the flagellum evolved is currently beyond us, an outline is known, including how some of the components might have been used originally for other purposes and co-opted to make the flagellum motor.

The study of biology makes it clear that living organisms are contraptions, cobbled together from odds and ends as circumstances dictate. Although many bits operate beautifully, a lot of the 'design' has a make-do air about it. In evolution, it is sufficient that organisms get by, they don't all have to be biological Rolls-Royces. Many features of the human body contain design flaws, such as the dangerous convergence of food and air pathways in the throat, and the inadequate robustness of the spine. If there is a designer, then this being is clearly not micro-managing the process very well.

The weak point in the 'gaps' argument of the Intelligent Design movement is that there is no reason why biologists should immediately have all the answers anyway. Just because something can't be explained in detail at this particular time doesn't mean that it has no natural explanation: it's just that we don't know what it is yet. Life is very complicated, and unravelling the minutiae of the evolutionary story in detail is an immense undertaking. Actually, in some cases we may never know the full story. Because evolution is a process which operates over billions of years, it is entirely likely that the records of many design-like features have been completely erased. But that is no excuse for invoking magic to fill in the gaps.[8]

One of the confusions surrounding the Intelligent Design movement's propaganda is a failure to distinguish between the *fact* of evolution and the *mechanism* of evolution. Design proponents often cite squabbles among biologists as signs that 'Darwinism is in trouble'. Even if it was, that doesn't mean that life hasn't evolved over billions of years. Darwinism proposes a specific physical mechanism. Other possible mechanisms may exist to drive evolution. For example, Jean-Baptiste Lamarck offered a theory of evolution based on the idea of inheriting acquired characteristics. That is to say, an organism's lifetime experiences can, according to this theory, be passed on to its offspring; so, for example, the son of a man who becomes a bodybuilding fanatic should have bigger than average muscles. Sadly for Lamarck, his theory has been refuted – more or less. The point, however, is that the theory has clearly defined and testable consequences, which qualifies it as being a *scientific* theory; the same cannot be said for Intelligent Design. Those who think that alternatives to Darwinism should be taught in schools would do well to consider Lamarck's theory of evolution for this purpose. It is not inconceivable (and it is certainly scientifically possible) that some subtle version of Lamarckian evolution might be at work here and there, augmenting the Darwinian mechanism.

Another possible evolutionary mechanism is self-organization. Many non-living systems evolve complex patterns and organizational structure from featureless beginnings. They do this entirely spontaneously, without variation or selection in the Darwinian sense. For example, snowflakes form distinctive elaborate hexagonal patterns. Nobody suggests that there are genes for a snowflake, but nobody suggests that they are made by an intelligent designer either. They spontaneously self-organize and self-assemble, in accordance with definite mathematical rules and physical laws. Non-Darwinian evolution by self-organization is found in physics, chemistry, astronomy, earth sciences and even in networks such as the World Wide Web. It would be surprising if it didn't happen here and there in biology too, but I may be wrong.[9] Even if I am right, that doesn't mean that Darwinism is disproved, just that it may be only a partial account of the evolutionary mechanism. But the missing bit is not some cosmic magician, it is a natural process complying with some yet to be elucidated principle of organization deriving from physical laws.

Further confusion in the intelligent design discussion often arises from a failure to distinguish between the *evolution* of life and the *emergence* of life – how life got started in the first place. Darwin himself pointedly omitted any reference to life's origin: 'One might as well speculate about the origin of matter,' he said (a problem which, incidentally, has now effectively been solved – see p. 120). It has to be admitted that the origin of life remains a deep mystery. But that cannot be used as an argument against Darwinian evolution, because biogenesis is not part of evolutionary theory. Clearly, we can discuss the evolution of life only on the basis that life already exists. So could it be that life's murky beginning is one of those 'irreducible' gaps in which the actions of an intelligent designer might lie? I don't think so. Let me repeat my warning. Just because we can't explain how life began doesn't make it a miracle. Nor does it mean that we will never be able to explain it, just that it's a hard and complicated problem about an event that happened a long time ago and left no known trace. But I for one am confident that we will figure out how it happened in the not too distant future.

In spite of the curious resurgence of the God-of-the-gaps argument in the United States, it remains true that Paley's biological design argument for the existence of God is blown out of the water by Darwin's theory of evolution and its refinements. But what about Paley's astronomical arguments? Here the situation is much more subtle.

Laws by design versus anthropic selection in a multiverse

In astronomy and cosmology, the appearance of design enters most strikingly when it comes to the laws of physics and the overall organization of the universe, especially in relation to the fine tuning and bio-friendliness I have been discussing throughout this book. Here the design argument is largely immune to Darwinian attack. The

Darwinian mechanism of variation, inheritance and selection cannot easily be adapted to cosmology.[10] There is no battle for survival, with universes slugging it out, red in tooth and claw, passing on their successful traits to baby universes, and no competition for resources or 'universe-eats-universe' struggle. It could be said that in the eternal inflation version of the multiverse, victory goes to the empty, featureless regions between the bubble universes, where inflation hasn't ceased – if by 'victory' one means gaining the biggest volume of space. But clearly, Darwinism is not an appropriate framework for explaining the appearance of design in cosmology.

There is, however, the possibility of anthropic selection rather than Darwinian selection. In the previous chapter I discussed how the multiverse theory combined with anthropic selection constituted a serious attempt to explain the appearance of design. This multiverse/anthropic challenge to intelligent design has already attracted the attention of the Roman Catholic Church. Christoph Schönborn, the Cardinal of Vienna, recently wrote in the *New York Times*:

Now at the beginning of the 21st century, faced with scientific claims like neo-Darwinism and the multiverse hypothesis in cosmology invented to avoid the overwhelming evidence for purpose and design found in modern science, the Catholic Church will again defend human reason by proclaiming that the immanent design evident in nature is real. Scientific theories that try to explain away the appearance of design as the result of 'chance and necessity' are not scientific at all, but, as [Pope John Paul II] put it, an abdication of human intelligence.[11]

The take-home message from the cardinal is that invoking a multiverse is an attempt to 'explain away' design rather than explain it, and that God is a better, simpler and more credible explanation.

So let's take a look at the hypothesis that the appearance of design in the universe is the result of a designer/creator. Although by definition this is not a scientific explanation (since it appeals to a supernatural cause), it is still a rational explanation. The question is, how good is it? We can crudely caricature an intelligent designer as a being who contemplates a shopping list of possible universes, figures out one that will contain life and observers, and then sets to work creating it, discarding the alternatives. There is no doubt that, even in this crude form, the hypothesis of an intelligent designer *applied to the laws of nature* is far superior than the designer considered in the previous section, who *violates* the laws of nature from time to time by working miracles in evolutionary history. Design-by-laws is incomparably more intelligent than design-by-miracles. If I were an omnipotent being who wanted to make an inhabited universe like ours, and I could achieve this simply by conjuring up what I wanted when I wanted it, I wouldn't regard my activities as very clever. But to select a set of laws that, without any periodic fixing up and micromanagement, can bring a universe into being and bring about self-organization, self-complexification and self-assembly of life and consciousness – well, that looks very clever indeed! So the 'intelligent' design beloved of the Intelligent Design movement strikes me as not very intelligent at all, in contrast to a designer of laws of nature which by themselves have such astonishing creative ability *without* the need for intervention and miracles.

Intelligent design *of the laws* does not conflict with science, because it accepts that the whole universe runs itself according to physical laws, and that everything that happens in the universe has a natural explanation. There are no miracles other than the miracle of nature itself. You don't even need a miracle to bring the universe into existence in the first place, because the big bang may be brought within the scope of physical laws too, either by using quantum cosmology to explain the origin of the universe from nothing, or by assuming something like eternal inflation.

The designer-of-laws is *responsible* for the universe, and might be thought of as upholding its existence at every moment, but does not tinker with its day to day operation. The type of God I am describing comes close, I think, to what many scholarly theologians – and for that matter quite a few scientists – profess to believe in. Yet even this 'no miracles' version of an intelligent designer is not without its critics. The central objection to invoking such a being to account for the ingenious form of the universe is the completely ad hoc nature of the explanation. Unless there is already some other reason to believe in the existence of the Great Designer, then merely declaring 'God did it!' tells us nothing at all. It simply plugs one gap – the mystery of cosmic bio-friendliness – with another – the mystery of an unknown intelligent designer. So we are no further forward.

A cosmic designer must lie outside time

There is also the very considerable problem of time. Time is part of the physical universe, inseparable from space and matter. Any designer/ creator of the universe must therefore transcend time, as well as space and matter. That is, God must lie *outside* time if God is to be the designer and creator *of* time. Augustine was well aware of this, and began a school of thought which asserts that God is a timeless being, not just in the sense of living for ever but of being outside time altogether.[12] (As I have explained, time itself can have a beginning and an end. Most theologians would not want God to come into being at the beginning of time and cease to exist if it should come to an end.) The difficulty with a timeless designer, however, is in making sense of the concept of design. What does it mean to design something 'timelessly'? In human experience, a designer is a being who thinks through in advance the consequences of certain choices, and then selects a judicious one. But 'thinking' and 'in advance' are inescapably temporal descriptions.

Even if some more abstract notion of 'timeless design' is accepted, a further difficulty arises with the specifics of the designer's choice. Could the designer have chosen a different universe, or chosen not to make a universe at all? If the answer is no, then God had no alternative but to create *this* universe and plays no role at all in the explanation – and so does not merit the title 'designer'. Nature is reduced to a subset of the divine being rather than a creation of this being – in fact, we might as well do away with the notion of a designer altogether. Christians, however, traditionally believe something quite different. They believe that God created this particular universe as a free act: that is, God was free to *not* make this universe. But this comes with its own set of difficulties, because we can ask why it was that God chose to make *this* universe, as opposed to a lifeless one, or one with maximum suffering. If the reply is, 'it's unfathomable', then the chain of explanation peters out. If the answer is that the choice was blind, then again the element of design is lost, because if the selection was purely whimsical, then the universe is reduced to a divine plaything. But if the answer is that the decision to make the universe was a profound and considered one which proceeded from God's nature, then one is prompted to ask about the source of this nature. In other words, who designed the designer? This is a variant of the old 'who made God?' conundrum.

Speculations (some of them wild) about a natural god

One way of evading the creator's creator difficulty comes from an imaginative suggestion by the cosmologist Edward Harrison, made in the context of the multiverse theory.[13] If one accepts that there are many universes, and that universes can be created by natural processes with different laws, constants and initial conditions, it is but a small step to the speculation that our universe is the engineered product of an intelligent designer who evolved naturally in an earlier universe. Harrison envisages a random ensemble of universes in which some pocket universes give rise to life and intelligence purely by chance. One of these universes develops a super-intelligence so technologically advanced that it becomes capable of creating baby universes to order (e.g. by gaining control over the universe-generating mechanism). These baby universes are deliberately made to optimize life and observers. Our universe would then be the product of a natural god who evolved by good old Darwinian processes in a preceding universe. Such a being has a long tradition in the history of religious philosophy, and is usually referred to as a demiurge. Plato's demiurge was a powerful creator of the world we see, but nevertheless had to work within the resources and laws available. The demiurge is not omnipotent, as is the monotheistic God. Harrison's god is, however, a super-demiurge because it can choose both the form of matter *and* the laws of low-energy physics by, for example, nucleating a universe in the appropriate region of the string theory landscape.[14] Nevertheless, this god is still bound by the laws of string/M theory (or whatever unified theory one envisages) and the physics of the universe-generating mechanism.

Harrison's speculation carries echoes of Hoyle's 'super-intelligence' who has deliberately 'monkeyed with the laws of physics', and of the Star Maker in Olaf Stapledon's famous novel.[15] Hoyle disliked the traditional Christian deity who makes the universe as a free choice, because it implies a lopsidedness of creator and creation: creation depends on God, but God remains unaffected by creation. If God has no need to create the world, and is unaffected by it, why did God bother to do so? This 'logical morass is avoided', Hoyle points out, if God exists 'only by virtue of the support received from the Universe'.[16] By this, Hoyle meant that God exists within the universe (or multiverse) rather than transcending it.

Similar ideas have been espoused by Andrei Linde in a curious paper informally entitled 'The hard art of universe creation' in which he discusses a super-civilization manipulating temperature, pressure and external fields so as to configure the birth of a universe with life-encouraging low-energy physics, 'to send a message to those who will live in the universe'.[17] The 'message in a bottle' universe was also a theme taken up by Heinz Pagels, who wondered whether the ingenious, life-encouraging laws of physics – what he termed 'the cosmic code' – might be a message from a demiurge:

Scientists in discovering this code are deciphering the Demiurge's hidden message, the tricks he used in creating the universe. No human mind could have arranged for any message so flawlessly coherent, so strangely imaginative, and sometimes downright bizarre. It must be the work of an Alien Intelligence.[18]

Pagels took pains to point out that his demiurge, his 'Alien Intelligence', was just an interesting thought experiment, and that the said being had somehow 'written himself out of the code'. The science writer James Gardner has adapted the same general concept into what he terms 'the selfish biocosm'. His thesis is that the universe is a self-organizing, self-replicating system in which life and intelligence emerge to create new universes with life and intelligence, with the 'knobs of the designer machine' suitably fine-tuned. 'Under the theory,' he writes, 'the laws and constants are life-friendly precisely because they were deliberately engineered by advanced intelligent life-forms in a prior cosmic cycle to endow our universe with the capability for life-mediated reproduction.'[19]

It is now time to take a reality check. In our search for an explanation of cosmic bio-friendliness we have encountered a heady mix of speculation, ranging from the intriguing to the seriously flaky. Some of the ideas have included fake universes designed by fake gods, multiverses which feature absolutely everything conceivable, and now super-intelligent, god-like beings who evolve naturally but then go on to create or manipulate entire universes for their own purposes. All these fanciful theories provide excellent material for entertaining science fiction, but the realm of professional science has been left far behind. If it were not for the fact that the speculators include some scientists of great distinction, the discussion could probably be dismissed without further ado. The fact that some great minds have been driven to explore such wild ideas is testimony to the intractable nature of the problems being confronted. Somehow we have to understand how life and cosmology connect (unless we are to dismiss the link as illusory). But even if we are prepared to suspend disbelief and go along with a natural god as a working hypothesis, the job is only half done.

God as a necessary being

A major shortcoming of invoking demiurges and natural gods is that they do not address the *ultimate* explanation of existence. One has to assume that some sort of universe or multiverse already exists before the god or gods can emerge. Professional theologians are unlikely to be impressed by that. They argue for the existence of a transcendent eternal deity who is outside, and responsible for, *all* universes at *all* times. Theologians have confronted the 'who-made-God' argument for centuries and have had plenty of time to come up with interesting answers. The conventional Christian doctrine is that God had no maker. Instead, God is a *necessary being* – a being whose existence needs no explanation in terms of something outside itself. In other words, it is logically impossible for God not to exist; a state of 'no God' is deemed to be meaningless.

It is far from clear to me whether such a conclusion is logically valid or conceptually coherent (even professional philosophers continue to squabble about it[20]), but even if it is, we are not done. Christians, like all monotheists, believe in *one* God. So they need to show not only that God exists necessarily, but that this being is necessarily unique – otherwise there could be countless necessary beings making countless universes. Even if all of this can be sorted out, we are still confronted with the problem that, in spite of God's necessary existence and nature, God did *not necessarily* create the universe as it is, but instead merely *chose* to do so. But now the alarm bells ring. Can a necessary being act in a manner that is not necessary?[21] Does that make sense? On the face of it, it doesn't. If God is necessarily as God is, then God's choices are necessarily as they are, and the freedom of choice evaporates.[22] Nevertheless, there is a long history of attempts to get round this obstacle and to reconcile a necessary God with a contingent universe.[23]

Confused? I certainly am. I am not an accomplished enough philosopher to evaluate these explanations, which become very technical. The arguments are abstract, subtle and convoluted, and the question inevitably arises of whether something like the multiverse/anthropic explanation isn't easier to grasp and altogether more plausible.

If only that were true.

Who designed the multiverse?

If the concept of 'God' runs into a logical and existential quagmire, then the multiverse fares little better. Just as one can mischievously ask who made God, or who designed the designer, so one can equally well ask why the multiverse exists and who or what designed it. Although a strong motivation for introducing the multiverse concept is to get rid of the need for design, this bid is only partially successful. Like the proverbial bump in the carpet, the popular multiverse models merely shift the problem elsewhere – up a level from universe to multiverse. To appreciate this, one only has to list the many assumptions that underpin the multiverse theory.

First, there has to be a universe-generating mechanism, such as eternal inflation. This mechanism is supposed to involve a natural, law-like process – in the case of eternal inflation, a quantum 'nucleation' of pocket universes, to be precise. But that raises the obvious question of the source of the quantum laws (not to mention the laws of gravitation, including the causal structure of spacetime on which those laws depend) that permit inflation. In the standard multiverse theory, the universe-generating laws are just accepted as given: they don't come out of the multiverse theory. Second, one has to assume that although different pocket universes have different laws, perhaps distributed randomly, nevertheless laws of some sort exist in every universe. Moreover, these laws are very specific in form: they are described by mathematical equations (as opposed to, say, ethical or aesthetic principles). Indeed, the entire subject is based on the assumption that the multiverse can be captured by (a rather restricted subset of) mathematics.

Furthermore, if we accept that the multiverse is predicted by string/M theory, then that theory, with its specific mathematical form, also has to be accepted as given – as existing without need for explanation. One could imagine a different unified theory – N theory, say – also with a dense landscape of possibilities. There is no limit to the number of possible unified theories one could concoct: O theory, P theory, Q theory . . . Yet one of these is assumed to be 'the right one' – without explanation. Now it may be argued that a decent theory of everything would spring from some deeper level of reasoning, containing natural and elegant mathematical objects which already commend themselves to pure mathematicians for their exquisite properties. It would – dare one say it? – display a sense of ingenious design. (Certainly the theoretical physicists who construct such theories consider their work to be designed with ingenuity.) In the past, mathematical beauty and depth have been a reliable guide to truth. Physicists have been drawn to elegant mathematical relationships which bind the subject together with economy and style, melding disparate qualities in subtle and harmonious ways. But this is to import a new factor into the argument – questions of aesthetics and taste. We are then on shaky ground indeed. It may be that M theory looks beautiful to its creators, but ugly to N theorists, who think that their theory is the most elegant. But then the O theorists disagree with both groups . . .

If there were a unique final theory,
God would be redundant

Let me now turn to the main scientific alternative to the multiverse: the possible existence of a *unique* final theory of everything, a theory that permits only *one* universe.[24] Remember that many scientists, including some prominent string theorists such as David Gross, are scathing about the multiverse, regarding those who espouse it as 'giving up'. They are convinced that a unique theory describing a unique world, with all laws and parameters completely fixed by the theory, will eventually emerge – maybe one day soon. Einstein once remarked that what interested him most was whether 'God had any choice in the creation of the world'. If Gross is right, the answer is no: the universe *has* to be as it is. There is only one mathematically self-consistent universe possible. And if there were no choice, then there need be no Chooser. God would have nothing to do because the universe would necessarily be as it is.

Intriguing though the idea of a 'no-free-parameters' theory may seem, there is a snag. If it were correct it would leave the peculiar bio-friendliness of the universe hanging as a complete coincidence. Here is a hypothetical unique theory which just happens, obligingly, to permit life and mind. How very convenient! To get some idea of what we are being invited to believe, consider an analogy.[25] The number π, defined as the ratio of the circumference to the diameter of a circle, was discovered by geometers in ancient Greece. It arises, however, in many contexts in the natural world, ranging from the motions of the planets to the patterns made by waves. One might say that π is built into the very structure of the physical universe. Now π cannot be expressed as a ratio of whole numbers (it is 'irrational', to use the technical term); rather, it must be expanded as an infinite decimal: 3.141 592 . . . without any end to the sequence of digits. It can also be expressed as an unending binary sequence, in terms of 1's and 0's. These strings of numbers, both decimal and binary, appear from statistical tests to be completely random. Knowing the first million digits of π gives no clue at all to the million-and-first digit.

Imagine next that the unending binary expression of π is displayed on a computer monitor in simple pictorial form, by making 1 corre-spond to a bright pixel and 0 to a dark pixel, starting with the first digit and continuing indefinitely. Being random, the sequence will generate screen after screen of uninteresting 'snow'. It is, however, a feature of randomness that, sooner or later, anything that is possible will happen. In this case, the implication is that somewhere deep in the binary expansion of π there will be a screen containing a coherent shape – a circle, say. It is easy to work out that the probability of this happening is so low that a computer could display a fresh screen showing the continuing binary expansion of π every second for a lifetime and you would still have a negligible chance of seeing a circle. Nevertheless, there is a non-zero probability that it could happen. The same logic applies to more complicated shapes, such as pictures of faces, but the more elaborate the image, the lower the probability of ever seeing it.

Now imagine that this experiment is performed, and after only two minutes a smiling face appears on the screen! What would one make of it? To say that one would be astonished is an understatement; some sort of trickery would be very strongly suspected. And yet the binary expression of π is uniquely fixed by the rules of mathematics. There are no free parameters to fiddle with to 'make a face'. If a face is there, it is there by the iron logic of real numbers: we need provide no further explanation. The same is true of the unique, no-free-parameters theory of everything, which, purely as a fortuitous consequence of its internal mathematical logic, describes a life-permitting universe (if this theory indeed exists). Of course, if you don't think that life and mind are anything very special, the digits of π analogy won't cut much ice. But in my opinion, life and mind *are* special, for reasons I shall explain in the next chapter. It seems to me that a unique mathematical theory that makes no reference to life, but which nevertheless yields life, is as unbelievable as seeing a face leap out from among the early digits of π.

A *unique final theory seems to be falsified already*

There is another, more direct argument against the idea of a unique final theory. The job of the theoretical physicist is to construct possible mathematical models of the world. These are often what are playfully called toy models: clearly too far removed from reality to qualify as serious descriptions of nature. Physicists construct them sometimes as an experiment, to test the consistency of certain mathematical techniques, but usually because the toy model accurately captures some limited aspect of the real world in spite of being hopelessly inadequate about the rest. The attraction is that such slimmed-down world models may be easy to explore mathematically, and the solutions can be a useful guide to the real world, even if the model is obviously unrealistic overall.

A good example of a toy model is one used to solve a problem in fewer than the usual three dimensions of space. I did a lot of this myself in the 1970s. I was interested in the behaviour of quantum fields propagating in curved spacetime, and it was sometimes possible to solve the equations exactly by pretending that space had only one dimension rather than three. For some real-world situations the two extra dimensions I omitted play an unimportant role, so my one-dimensional calculations provided a useful guide. Such toy models are a description, not of the real world but of impoverished alternatives. Nevertheless, they describe *possible* worlds. Anyone who wanted to argue that there can be only one truly self-consistent theory of the universe would have to give a reason why these countless mathematical models that populate the pages of theoretical physics and mathematics journals were somehow unacceptable descriptions of logically possible worlds.[26]

It's not necessary to consider radically different universes to make the foregoing point. Let's start with the universe as we know it, and change something by fiat: for example, make the electron heavier and leave everything else alone. Would this arrangement not describe a possible universe, yet one that is different from our universe? 'Hold on,' cries the no-free-parameters proponent, 'you can't just fix the constants of nature willy-nilly and declare that you have a theory of everything! There is much more to a theory than a dry list of numbers. There has to be a unifying mathematical framework from which these numbers emerge as only a small part of the story.' That is true. But I can always fit a finite set of parameters to a limitless number of mathematical structures, by trial and error if necessary. Of course, these mathematical structures may well be ugly and complicated, but that is an aesthetic judgement, not a logical one. So there is clearly no unique theory of everything if one is prepared to entertain other possible universes and ugly mathematics.

While the argument I have just given seems to be incontestable, many physicists would be content to settle for a weaker claim than the assertion that the universe cannot possibly have been other than it is. To be sure, there may exist legions of self-consistent mathematical 'theories of everything' describing universes different from ours, but when it comes to *this* universe, perhaps there is only one self-consistent theory. Maybe if we knew enough about unifying theories we would find that only *one* knob setting of the Designer Machine (i.e. only one theory) fits all the known facts about the world – not just the values of the constants of nature, but such things as the existence of life and observers. That would certainly be something! It could be that there are many possible no-free-parameter theories describing many possible single universes, but only one of those theories *fits all the observed facts about this universe*. Surely this imaginary Holy Grail of a theory solves the Goldilocks enigma without invoking either intelligent design or a multiverse?

Unfortunately it would not, because one would still be left with the puzzle of why *that* theory – the one that permits a life-giving universe – is 'the chosen one'. Stephen Hawking has expressed this more eloquently: 'What is it that breathes fire into the equations and makes a universe for them to describe?'[27] Who, or what, does the choosing? Who, or what, promotes the 'merely possible' to the 'actually existing'? This question is the analogue of the problem of 'who made God' or 'who designed the Designer'. We still have to accept as 'given', without explanation, one particular theory, one specific mathematical description, drawn from a limitless number of possibilities. And the universes described by almost all the other theories would be barren.

Perhaps there is no reason at all why 'the chosen one' is chosen. Perhaps it is arbitrary. If so, we are left still with the Goldilocks puzzle. What are the chances that a randomly chosen theory of everything would describe a life-permitting universe? Negligible. There are many locked knob-settings on the Designer Machine representing logically possible single, no-free-parameter, mathematically ugly, sterile universes. If any one of these infinitely many possibilities had been the one to 'have fire breathed into it' (by a Designer with poor taste perhaps?), we wouldn't know about it because it would have gone unobserved and uncelebrated. So it remains a complete mystery as to why *this* universe, with life and mind, is 'the one'.[28]

My conclusion is that both the multiverse theory and the putative no-free-parameters theory might go a long way to explaining the nature of the physical universe, but nevertheless they would not, and cannot, provide a complete and final explanation of why the universe is fit for life, or why it exists at all.

What exists and what doesn't: who or what gets to decide?

We have now reached the core of this entire discussion, the problem that has tantalized philosophers, theologians and scientists for millennia:

What is it that determines what exists?

Among the big questions I listed at the beginning of this book, this is one of the Really Big ones. The physical world contains certain objects – stars, planets, atoms, living organisms, for example. Why do *those* things exist rather than others? Why isn't the universe filled with, say, pulsating green jelly, or interwoven chains, or disembodied thoughts . . . ? The possibilities are limited only by our imagination. The same sort of conundrum arises when we contemplate the laws of physics. Why does gravity obey an inverse square law rather than, for example, an inverse cubed law? Why are there two varieties of electric charge (+ and −) instead of four? And so on. Invoking a multiverse merely pushes the problem back to 'why *that* multiverse'. Resorting to a no-free-parameters single universe described by a unified theory invites the retort 'Why *that* theory?'

Is there a way out? There is, but it is pretty drastic. There are only two of what one might term 'natural' states of affairs, by which I mean states of affairs that require no additional justification, no Chooser and no Designer, and are not arbitrary and reasonless. The first is that *nothing* exists. This state of affairs is certainly simple, and I suppose it could be described as elegant in an austere sort of way, but it is clearly wrong. We can confidently rule it out by observation. The second natural state of affairs is that *everything* exists. By this I mean that everything that *can* exist *does* exist. Now that contention is much harder to knock down. We can't observe everything in the universe, and absence of evidence is not the same as evidence of absence. We cannot be sure that any particular thing we might care to imagine[29] doesn't exist *somewhere*, perhaps beyond the reach of our most powerful instruments, or in some parallel universe.

Could it be that everything exists?

An enthusiastic proponent of this extravagant hypothesis is Max Tegmark.[30] He was contemplating the 'fire-breathing' conundrum I discussed above (allegedly over a few beers in a pub). 'If the universe is inherently mathematical, then why was only one of the many mathematical structures singled out to describe a universe?' he wondered. 'A fundamental asymmetry appears to be built into the heart of reality.' To restore the symmetry completely, and eliminate the need for a Cosmic Selector, Tegmark proposed that 'every mathematical structure corresponds to a parallel universe'. So this is a multiverse with a vengeance. On top of the 'standard' multiverse I have already described, consisting of other bubbles in space with other laws of physics, there would be much more: 'The elements of this [extended] multiverse do not reside in the same space but exist outside of space and time. Most of them are probably devoid of observers.'[31]

Tegmark gives an illustration of the other mathematical structures he has in mind. 'How about time that comes in discrete steps, as for computers, instead of being continuous? How about a universe that is simply an empty dodecahedron?'[32] He also considers fractals – mathematical structures that have fractional dimensionality (e.g. dimension $3^1/_3$), and are infinitely irregular.[33] There would also be sparse universes that would be just sets of points with no connectivity, others containing objects that could be counted, but for which 3×4 would not be the same as 4×3. Some universes might allow for 'hypercomputation' – the ability to solve mathematical problems that would take an infinite number of steps in our universe. Mathematicians consider these sorts of systems all the time, and write papers about them as branches of mathematical theory. Tegmark suggests that they also correspond to physical reality – somewhere.

A knee-jerk reaction to Tegmark's 'anything goes' multiverse[34] is that it is stupefyingly complex and in flagrant breach of the most basic rule of science, which is Occam's razor. But Tegmark points out that everything is actually simpler than something. That is, the whole can often be defined more economically than any of its parts. A concrete example is provided by an infinite crystal. A perfect infinite crystal consists of a regular and uniform lattice of atoms without any boundary. Its structure may be described completely by simply specifying the spacing between neighbouring atoms, the overall orientation, and the statement that the lattice has no boundary. Now imagine removing from the perfect infinite crystal a random set of atoms and separately retaining the three-dimensional arrangements of both the removed set and the remainder. In this manner the original set of atoms has been divided into two randomly specified subsets. Now, by definition a random set cannot be described by a smaller number of bits of information than the set itself contains.[35] For example, if you remove one million atoms at random you must specify one million bits of information to identify the process. So each subset is a complicated object needing a lot of information to describe it. But put the two subsets back together again and you get something very simple and easy to describe.

Beguiling though Tegmark's proposal may seem, it is not without problems. The notion of 'everything' is a slippery one in mathematics, because of the possibility of self-referential sets. These relate to the well-known problems that lurk within the logical foundations of mathematics, such as Bertrand Russell's barber paradox. The village barber shaves all the men who don't shave themselves. Then who shaves the barber? If the barber shaves himself, he belongs to the set of men who are *not* shaved by the barber, so he doesn't shave himself. But if he doesn't shave himself, then he *is* shaved by the barber! Either way, one gets contradictory nonsense. This little puzzle is more than just an amusing diversion: it strikes at the very roots of mathematical consistency, and negates any simple attempt to define mathematics in terms of sets of objects obeying sets of rules. Even basic arithmetic is compromised.[36]

A second objection to Tegmark's idea is that it is framed in terms of mathematics. This is, of course, hardly surprising of a theory constructed by a mathematical physicist, and it is supported by the discovery that nature is indeed mathematical in form. But if we are in 'anything goes' territory, there is no reason at all to restrict ourselves to mathematics. To do so invites the question of *why* the multiverse is mathematical: who decided *that*? We can certainly imagine multiverses that are defined in other ways. For example, the set of all aesthetically pleasing universes, the set of all good universes, or evil universes, the set of all deities, the set of all virtual realities. Even within the general domain of mathematics, one can question the standard rules of logic on which mathematical statements are based, and contemplate a multiverse with the set of all objects deriving from all possible forms of logic. If you really believe that everything exists, then all these other multiverses should be out there too.

Maybe they are? Very few people, however, would want to go the whole hog with Tegmark. Most scientists, even those who believe in some sort of multiverse, stop short of supposing that literally everything exists. Which brings us right back to the basic problem we confronted at the outset of this section: what decides what exists?

The origin of the rule that separates what exists from what doesn't

If not everything exists, there must be a description or a rule that specifies how to separate 'the actual' from 'the possible-but-in-fact-non-existent'. The inevitable questions then arise: What is the rule that divides them? What, exactly, determines that-which-exists and separates it from that-which-might-have-existed-but-doesn't? From the bottomless pit of possible entities, something plucks out a subset and bestows upon its members the privilege of existing. Something 'breathes fire into the equations' and makes a universe or a multiverse for them to describe. And the puzzle doesn't stop there. Not only do we need to identify a 'fire-breathing actualizer' to promote the merely-possible to the actually-existing, we need to think about the origin of the rule itself – the rule that decides what gets fire breathed into it and what doesn't. Where did that rule come from? And why does *that* rule apply rather than some other rule? In short, how did the right stuff get selected? Are we not back with some version of a Designer/Creator/Selector entity, a necessary being who chooses 'the Rule' and 'breathes fire' into it?

These deliberations are summarized in Figure 26, which shows three sets separated by two boundaries, A and B. The darkest region, enclosed within boundary A, is the set of all things that all possible observers can in principle observe. This set can be a part – a subset – of all that exists, which is encompassed by boundary B. Excluded from the darkest subset, but still contained within boundary B, are, for example, universes which contain no life and observers. The intermediate shaded region on the other side of boundary B is the set of things that could in principle have existed, but in fact do not exist. The union of these three sets forms the set of all that is logically possible – that is, all that could in principle exist.

Let's see how the three attempts to explain the universe relate to these sets. First consider the no-free-parameters theory of everything. Its advocates declare, quite simply, that boundary A does not exist: that there is just one universe – the *observed* universe (see Figure 27). They have no truck with notion of 'other universes' which go unseen

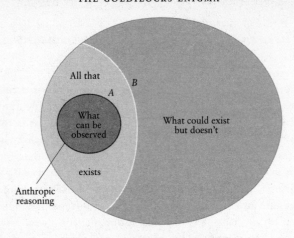

26. *Existence in a nutshell*

A schematic depiction of the set of all that can in principle exist. A distinction is made between what *actually* exists, and what *could* exist but doesn't. Included within the former set is a smaller subset, consisting of all that can in principle be observed. Left out of this subset are other things that exist but are unobservable (e.g. universes that do not permit life). Anthropic reasoning may be applied to boundary *A*, to explain why we observe a life-friendly universe, but not to boundary *B* (to determine the rule that separates what *can* exist from what *does* exist).

on account of being lifeless, so *A* is abolished. Unfortunately there is still boundary *B*. The theory of everything has nothing to say about *B*: it is left as a mystery.

How does the multiverse theory fare? Its advocates agree that there *is* a boundary *A*, and they invoke the anthropic principle, or the observer selection effect, to explain its size and shape (whatever they might be – we don't know yet). Once again, however, this theory cannot help us chart the boundary of the second set, boundary *B* – the dividing line between the set of all that *does* exist and the set of all that *can* exist, but doesn't. Boundary *B* is beyond the reach of anthropic reasoning. The multiverse/anthropic theory can perhaps explain why the universe is bio-friendly, but it is absolutely not a complete theory of the universe, however much that claim may be

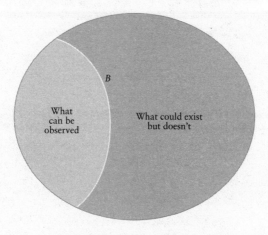

27. What if there is just one universe?

Like Figure 26, this diagram also depicts all that can in principle exist. Some physicists hope that there is only one possible universe consistent with a final, mathematically unified super-theory (e.g. M theory) and that what we observe is *it*. (It would have to be!) If they are right, then what gets observed is *all there is*. (We may not be able observe it all at any one time, but what we don't observe right now would be the same sort of thing as what we do observe.) However, this mathematically unique theory of everything would still not explain why *this* universe is the one that exists, as opposed to all other logically possible universes which don't 'have fire breathed into them', for example universes described by other unified theories that do not fit the observational facts of our universe, depicted here by the darker shaded region on the other side of boundary *B*.

made. It still leaves a lot unexplained. For one thing, we still need to know who, or what, drew boundary *B*. Now according to Tegmark's version of theory 2, boundary *B* does not exist: by his reasoning, the set of all that exists is the *same* as the set of all that can exist (see Figure 28). But how many people are really prepared to go that far? When it comes to the existence business, most people think that some things got left out. But what? And why *those* things?

The third theory is that some sort of divine selection operates. In crude terms, this theory says that God, or a set of gods, determines

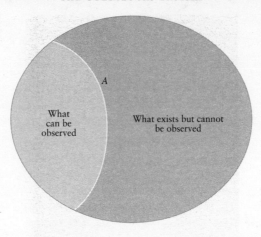

28. *Tegmark's 'anything goes' multiverse*

According to Max Tegmark, everything that can exist does exist: that is, all logically possible universes described by all possible mathematical structures are really out there somewhere. Anthropic reasoning may be used to define the (tiny!) subset of universes that can be observed (the lighter shaded region). The rest, located on the other side of boundary *A*, is the set of all possible things that can – and do – exist, and that go unobserved.

both boundaries *A* and *B* in Figure 26. In Christian theology, God either transcends the set of existing things or embeds it as a subset. But now there is a further mystery, and another boundary: the boundary between the actually existing God, and the set of all possible gods.[37]

Turtle power

There is a famous story (attributed by some to Bertrand Russell, by others to the nineteenth-century American philosopher William James), about a lecture on the nature of the universe. Part way through the talk a woman at the back stands up and denounces the lecturer, claiming that she *knows* how the universe is put together: the Earth rests on the back of a giant elephant which stands on the back of a giant turtle. The bewildered lecturer responds by asking what the

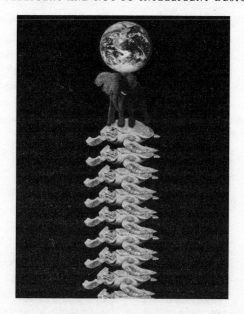

29. Turtles all the way down!
This humorous symbolism depicts an infinite regress in which the physical world of experience is explained by a deeper reality, which is in turn explained by a deeper reality, which . . .

turtle was standing on. 'You may be very clever, young man,' the woman shoots back, 'but you can't fool me. It's turtles all the way down!' (see Figure 29).

This light-hearted anecdote illustrates a seemingly unavoidable problem which confronts attempts to give a complete account of reality, and that is how to terminate the chain of explanation. In order to 'explain' something, in the everyday sense, you have to start somewhere. To avoid an infinite regress – a bottomless tower of turtles – you have at some point to accept *something* as 'given', something which other people can acknowledge as true without further justification. In proving a geometrical theorem, for example, one begins with the axioms of geometry,[38] which are accepted as self-evidently true and are then used to deduce the theorem in a step-by-step logical argument. Sticking to the herpetological metaphor, the axioms of

30. Super-turtle!
To avoid an infinite regress (the bottomless tower of turtles in Figure 29), one might consider a levitating super-turtle, which is self-explaining and self-supporting. Theologians call this 'a necessary being', and some have tried to prove that such a being exists. Some scientists have argued for the necessary existence of a unique super-unified theory.

geometry represent a levitating super-turtle, a turtle that holds itself up without the need for additional support (see Figure 30). The same general argument applies to the search for an ultimate explanation of physical existence.

The trouble is, one man's super-turtle is another man's laughing stock. Scientists who crave a theory of everything with no free parameters are happy to accept the equations of that theory (e.g. M theory) as their levitating super-turtle. *That* is their starting point. The equations must be accepted as 'given', and used as the unexplained foundation upon which an account of all physical existence is erected. Multiverse devotees (apart perhaps from Tegmark) accept a package of marvels, including a universe-generating mechanism, quantum mechanics, relativity and a host of other technical prerequisites as their super-turtle. Monotheistic theologians cast a necessary God in the role of super-turtle. All three camps denounce the other's super-turtles in equally derisory measure. But there can be no reasoned resolution of this debate, because at the end of the day one super-turtle or another has to be taken on faith (or at least provisionally accepted as a working hypothesis), and a decision about which one to pick will inevitably reflect the cultural prejudices of the devotee.[39] You can't use science to disprove the existence of a supernatural God, and you can't use religion to disprove the existence of self-supporting physical laws.

Many people think the universe is absurd

The root of the turtle trouble can be traced to the orthodox nature of reasoned argument. The entire scientific enterprise is predicated on the assumption that there are *reasons* for why things are as they are. A scientific explanation of a phenomenon is a rational argument that links the phenomenon to something deeper and simpler. That in turn may be linked to something yet deeper, and so on. For example, the changing shape of the moon from crescent to full and back again is explained by the motions of the bodies in the solar system, which are explained by Newton's laws, which are subsumed in Einstein's general theory of relativity, which (it is to be hoped) will emerge one day from a theory of quantum gravity such as string theory. Following the chain of explanation back (or the turtles down), we see that there are indeed rational reasons for why the moon changes shape, but when we reach the putative final theory – the super-turtle – what then? One can ask: Why *that* unified theory rather than some other? Why a unified theory that permits a moon? Why a unified theory that permits sentient beings who can observe the moon? One answer you may be given is that there *is* no reason: the unified theory must simply be treated as 'the right one',[40] and its consistency with the existence of a moon, or of living observers, is dismissed as an inconsequential fluke. If that is so, then the unified theory – the very basis for all physical reality – itself exists *for no reason at all*. Anything which exists reasonlessly is by definition absurd. So we are asked to accept that the mighty edifice of scientific rationality – indeed, the very mathematical order of the universe – is ultimately rooted in absurdity! There is no reason at all for the scientific super-turtle's amazing levitating power.

A different response to such questions comes from the multiverse theory. Its starting point is not a single, arbitrary set of monolithic laws, with fluky, unexplained bio-friendliness, but a vast array of laws, with the life factor accounted for by observer selection. But unless one opts for the Tegmark 'anything goes' extreme, then there is still an unexplained super-turtle in the guise of a particular form of multiverse based on a particular universe-generating mechanism and all the other paraphernalia. So the multiverse likewise retains an element of arbitrariness and absurdity. Its super-turtle also levitates for no reason, so that theory too is ultimately absurd.

Monotheistic theologians, for whom God plays the role of super-turtle, have had longer to think about this problem. They believe, or at least some do, that the threat of ultimate absurdity is countered by positing that God is a necessary being. As I explained earlier in this chapter, this is an attempt (and one which is not obviously successful) at describing a 'self-levitation' mechanism – God explains God's own existence – without which we would be right back to arbitrariness, reasonlessness and absurdity: if God exists reasonlessly, then the theistic explanation is also absurd. Whatever the dubiousness of the notion of a necessary being, at least the theologians have had a serious go at explaining *how* their super-turtle levitates rather than claiming that it happens for no reason.

The ultimate explanation of the universe should be simple

It would appear that each of the three positions I have discussed threatens to be ultimately absurd, and requires us to accept a starting point based on faith: either a set of mathematical laws, a multiverse with by-laws, or God. How are we to decide between them? At this point, 'the argument from simplicity' is often invoked. Martin

Gardner uses simplicity to argue for God over the multiverse: 'Surely the conjecture that there is just one universe and its Creator is infinitely simpler and easier to believe than that there are countless billions upon billions of worlds.'[41] The theologian Richard Swinburne has also argued that an infinite rational God is a simpler starting point than either an unexplained multiverse or a single ordered universe, and therefore constitutes a preferable starting point in the chain of reasoning.[42] However, Richard Dawkins has used the simplicity criterion to argue precisely the reverse, declaring that 'you can't get much more complex than an Almighty God!'[43] On the face of it, Dawkins has a point: an *infinite* mind (i.e. the traditional monotheistic God) looks to be infinitely *complex*, and not simple at all. But the same criticism can be levelled at an infinite multiverse, which requires an infinite amount of unverifiable information to specify it.[44] If it boils down to a choice between an infinite unseen God or an infinite collection of unseen universes, which explanation is the simpler? It seems to me that the traditional God of monotheism and the standard model of the multiverse are about equally complex in this regard. The hands-down winner in this 'simpler-than-thou' three-way contest would be the unique-universe, no-free-parameters theory – if such a theory exists, and if (it's a big if) it turns out to be the hoped-for *simple* and elegant description and not an unholy mishmash of complex mathematics.

It should be clear from this chapter that all three attempts to explain the world completely – two scientific and one theological – eventually hit the buffers, and demand that something truly huge be accepted on faith alone. So is that as far as we can go? Have we reached the final limits of reasoning, with no clear answer in sight? I don't think so. I set out in this book to address the question of why the universe is bio-friendly. I now want to take a deeper look at the 'bio' part. In the discussion so far, life and observers have been cast in a purely passive role. In the multiverse theory, for example, they exist in some universes and not in others, and therefore (without doing anything other than what they normally do – i.e. living and observing) they allegedly explain cosmic bio-friendliness. So long as life and mind are treated in this manner, as simply an incidental by-product of nature, then the mystery of existence will remain forever that – a mystery. But there is another avenue we can explore, one in which life and mind are not just carried along for the ride, but play an active role in explaining existence.

Key points

This chapter contains more of the author's personal opinions than the earlier chapters.

- In some respects, such as its strange fitness for life, the universe *appears* designed. Some people suggest that it *is*. Invoking a supernatural designer, however, opens up its own as-yet unsolved philosophical issues about the designer's nature, origin, uniqueness, necessity and relationship to time.
- It is important to distinguish between design in the laws of physics and design in objects or in systems such as biological organisms. The appearance of design in organisms has a testable scientific explanation based on Darwin's theory of evolution.
- The appearance of design in the laws might be explained by hypothesizing a multiverse with 'anthropic' selection. However, a multiverse is not a complete explanation of existence, because it still requires *some* unexplained physical laws.
- Some scientists pin their hopes on a complete 'theory of everything' that would explain the universe without invoking observer selection effects. If such a theory were correct, the bio-friendliness of the universe would be a lucky coincidence. However, the existence of a *unique* final theory is very doubtful.
- Unless everything that can exist does exist, something still unexplained must separate what exists from what doesn't.
- To avoid an infinite regress (a tower of turtles), something must be accepted on faith, even in scientific accounts, but that something should be as simple as possible.
- We are not finished yet!

10

How Come Existence?

Is life written into the laws of the universe?

A few years ago, Stephen Hawking summed up scientists' prevailing attitude to the status of life in the universe. 'The human race is just a chemical scum on a moderate-sized planet,' is how he put it.[1] Most physicists and cosmologists would echo Hawking, and regard life as a trivial, accidental embellishment to the physical world, of no particular significance in the overall cosmic scheme of things. But there has always existed a dissenting minority who maintain that life is not merely an incidental by-product of nature, but a deeply significant part of the cosmic story.

The anthropic principle seems to elevate life and mind to a special place in nature. However, in the form in which I have been discussing it so far – as a passive selection mechanism in a multiverse – it is no more than a necessary statistical procedure, usually referred to as the *weak anthropic principle*. When Brandon Carter introduced the term thirty years ago, he also explored the notion of a *strong anthropic principle*.[2] It was later elaborated by John Barrow and Frank Tipler.[3] Roughly speaking, the strong version of the principle asserts that the universe *must* be such as to give rise to observers at some stage in its development. In other words, the laws of physics and the evolution of the universe are in some unspecified manner destined to bring forth life and mind. The strong anthropic principle does not forbid a multiplicity of universes, but it stipulates that universes without life and observers don't exist. In this way, life in general, and conscious beings in particular, are treated as an active selection mechanism.

The same basic idea has been propounded, explicitly or implicitly, by many distinguished scientists. For example, Freeman Dyson: 'As we look out into the universe and identify the many accidents of physics and astronomy that have worked together to our benefit, it almost seems as if the universe must in some sense have known we were coming.'[4] Likewise the Cambridge biologist Simon Conway Morris says that, 'there is, if you like, seeded into the initiation of the universe itself the inevitability of intelligence'.[5]

The strong anthropic principle receives some support from the widespread belief that the emergence of life is somehow inevitable because it is 'built into' the laws of the universe. The Nobel prize-winning biologist Christian de Duve describes the universe as 'pregnant with life' and calls life 'a cosmic imperative'.[6] The biophysicist Stuart Kauffman echoes Freeman Dyson by declaring that we are 'at home in the universe'.[7] Behind the ambitious astrobiology programme funded by NASA and other institutions, and the SETI project to seek evidence of extraterrestrial intelligence, is the assumption that life is not a freak phenomenon confined to Earth, but a widespread and inevitable outcome of physical laws which are intrinsically slanted in favour of biology. In other words, life is not just a random by-product of nature, but a fundamental part of the workings of the cosmos. Such a view has an obvious popular appeal, but is it credible?

Taking life seriously

Scientists use the word 'fundamental' to describe something which goes to the very heart of nature and on which our broader understanding of the physical universe crucially depends. For example, electrons and quarks are called fundamental particles because they play a basic role in explaining the nature of matter without appealing to any deeper level of structure. Similarly, the force of gravity is fundamental because it serves to shape the structure of the universe. Space and time are fundamental for reasons that scarcely have to be spelt out. These examples may not describe the *most* fundamental entities in nature, but they are certainly more fundamental than, say, rain clouds or rocks or glue. These we might describe as merely incidental features of the world; their properties follow from those of the more fundamental entities, and while they may matter in daily life (will it rain, will my envelope come unstuck . . .) it would be perverse to ascribe cosmic significance to them. A truly fundamental entity cannot be described in terms of anything deeper, simpler or 'more' fundamental. Opinions differ, of course, about what might or might not be truly fundamental. Physicists used to think that electrons are truly fundamental, but today many of them would place strings in that deeper role, regarding electrons as just one manifestation of string activity. The question before us, then, is whether living organisms may legitimately be regarded as in some sense fundamental – and therefore significant – or incidental and insignificant side-effects of the main game.

Two hundred years ago, many scientists were content to treat life as a fundamental phenomenon because they believed that some sort of life force or vital essence was responsible for the remarkable qualities that living organisms display. This 'life stuff' was not supposed to be explained by anything deeper, but was accepted as a primitive, given property of biology. Today we know that there is no life force. Living organisms are machines, and they derive their extraordinary qualities from their great complexity.

What makes life special is not the stuff of which it is made, but the things it does. Defining life is notoriously hard, but three properties stand out. The first is that biological organisms are a product of Darwinian evolution; indeed, some scientists define life by that criterion alone. The evolutionary principle of replication with variation and selection is undeniably fundamental. It should apply to life everywhere in the universe, even forms of life very different from the terrestrial variety. Although Darwinian evolution is not a law of physics as such, it is an organizational principle as deep and significant as the law of gravity. Life is therefore a product of this very basic property of the universe.

The second key quality is autonomy. Living organisms literally have a life of their own. Although they are subject to the same physical forces as all other material systems, living organisms are able to *harness* those forces to carry out an agenda. A simple example will suffice to make this point obvious. If you throw a dead bird into the air, it will follow a simple geometrical path and land at a predictable spot. But if you throw a live bird into the air, it is impossible to know how it will move or where it will land. Crucially, the organism's unpredictability is quite distinct from chaotic or random behaviour, like the throw of a die or the fate of swirling eddies in a stream. The bird's trajectory is shaped in part by its genetic and neurological states.

The third distinctive property of living systems is how they handle information. All physical systems can be thought of as processing information in an elementary sense. For example, the position of a planet in space requires some numbers to specify it. As the planet orbits the sun, its position changes and the numbers will be different. So the simple process of planetary motion transforms 'input information' (the initial position of the planet) into 'output information' (the final position of the planet). But the information contained in a genome or a brain is more than mere data of this sort: a genome is a blueprint or algorithm or set of instructions for carrying out a project, for example making a protein, copying a molecule or seeking food. For an algorithm to work successfully, there has to be a physical system (in the case of a genome it could be a ribosome) that can interpret and carry out the genetic instructions. Those instructions 'mean something' to the system, which then 'acts' on them. Philosophers and computer scientists refer to meaningful information (as opposed to raw bits) as *semantic*. There is thus a semantic or contextual dimension to biological information: genetic information is not just a string of arbitrary bits, but a type of coherent computer program encoding a predefined goal, written in the four-letter alphabet of DNA. When it comes to consciousness, as opposed to mere biochemical activity, the semantic nature of neural information processing is obvious. Without doubt, minds process meaningful information: that is more or less what minds do.

The physicist David Deutsch, known for his pioneering work on quantum computation, takes the computational dimension of life beyond mere analogy. He points out that a genome contains an internal representation of the world – a type of virtual reality – constructed over aeons of evolution, incorporating the necessary contextual information for the associated organism to be well adapted to its environmental niche. In other words, the genome 'knows about' its environment. 'Now we are getting closer to the reason why life is fundamental,' he writes. 'Life is about the physical embodiment of knowledge. It says that it is possible to embody the laws of physics, as they apply to every physically possible environment, in programs for a virtual-reality generator. Genes are such programs.'[8] The ability of physical systems such as living organisms, brains and computers to construct a computational representation of the universe – to simulate it – is by no means a trivial property of the material world. It hinges on the so-called Turing principle (named after Alan Turing, co-inventor of the computer). Deutsch defines the Turing principle, which he considers to be on a par with the laws of physics as among the most fundamental properties of the physical world, as follows: 'It is possible to build a universal computer: a machine that can be programmed to perform any computation that any other physical object can perform.'[9] Although many people take this principle for granted in the computer age, it actually represents a very deep property of the world, and hinges on the types of material system that exist in nature and the way they behave.

By extension, Deutsch places *knowledge* alongside such things as mass and electric charge as a fundamental physical quantity, and he gives a curious argument to justify this claim. Imagine a future civilization on Earth with the technology to modify not just the planet (as we have already done, for good or bad) but the whole solar system, including the sun. Perhaps this civilization wishes to use its knowledge of astrophysics to prolong the lifetime of the sun, by altering its composition in some way. Now the evolution of stars like the sun is already well understood, and the properties of an ageing sun can be determined rather precisely by the application of the standard laws of nuclear and plasma physics. An alien observer on the far side of the galaxy who modelled the behaviour of our sun in this manner would fail to find agreement with observation, because the sun would have been altered by the scientific knowledge of the terrestrial civilization. In this case, knowledge has an impact big enough to rival the standard processes in astrophysics, such as the flow of heat from a stellar core. Of course, cosmic engineering is as yet the stuff of science fiction. But there is no reason in principle why life and mind cannot, over aeons, transform the structure of the universe on a very large scale. (This is a topic to which I shall return later in the chapter.) In any case, as Deutsch reminds us, size isn't everything. A phenomenon such as quantum interference (see p. 275) is unequivocally fundamental in character, yet exceedingly inconspicuous in almost all circumstances.[10]

Critics may object that knowledge is a human concept which has no place in any theory of the physical world, and especially not in cosmology. Can one really attribute 'knowledge' to a gene, say? Isn't it just a dumb molecule? It's true that segments of a molecule such as DNA don't come with a label 'I have knowledge' attached. To make this point obvious, consider the genetic database of life contained in the sequence of nucleotides, or letters, in DNA. These come in four varieties, denoted by A, T, C and G. When scientists analyse a gene to determine the sequence of nucleotides, the result is presented in the form of a long string of such letters, which might read in part AAGCTCGTTAGAC, for example. The primary job of genes is to code for the manufacture of proteins. Most DNA in complex organisms such as human beings is in fact 'non-coding' – it is not genetic information, and is often referred to as 'junk DNA'.[11] However, the *same* sequence of letters, AAGCTCGTTAGAC say, may occur both in a gene and in junk, and you can't tell simply by looking at the sequence on its own whether it 'has knowledge' (i.e. belongs to a gene) or hasn't (i.e. belongs to the junk). These molecular strands are physically identical. Only within the overall context of the organism and its environment does the knowledge designation make sense.

The trouble is, 'overall context' seems a very slippery, subjective concept. Deutsch, however, believes that he has a way to define it so as to give it objective significance. He points out that whereas changes in the coding, or genetic, part of DNA could be disastrous, by causing a mutation that would reduce the organism's adaptive success, most changes in the junk part of DNA seem to have no serious consequences. Natural selection therefore works to conserve the genetic sequences in DNA over many generations, but usually permits the junk sequences to drift randomly from one generation to the next. A mutation – that is, a change in the sequence of letters – may come about because a subatomic particle such as a cosmic ray traverses the DNA molecule and damages it, causing a rearrangement of the letters; just one letter out of place can have momentous consequences for the organism and its descendants. Such an event is highly sensitive to the

8. *The quantum multiverse*

Historically, the first theory of the multiverse was not primarily a cosmological idea at all, but an interpretation of quantum mechanics. Quantum weirdness implies that a particle such as an electron possesses an intrinsic uncertainty, quantified by Heisenberg's uncertainty principle. For example, when an electron scatters from a target we cannot know in advance whether it will bounce to the left or to the right. Quantum mechanics will give the relative odds for either outcome, but it cannot predict which will actually occur in any given case. However, we have no difficulty after the event in finding out which way the electron has scattered – we simply make a direct observation of its path. One way to think about this is to say that before the electron hits the target there is one world with two alternative futures: one with a left-moving electron, the other with a right-moving electron. At the moment of collision, nature is forced to make up its mind – left or right.

In the original interpretation of quantum theory, developed in Copenhagen in the 1930s under the influence of Niels Bohr, the transition from an uncertain, ghostly superposition of worlds to a single, concrete reality was attributed to the intervention of the experimenter. According to the Copenhagen interpretation, the act of observation itself was the key step in forcing nature to 'make up its mind' (left or right). A few physicists saw this as evidence for consciousness playing a direct role in the physical world at the quantum level. Most physicists, however, rejected that view. Although there is no consensus, one currently popular interpretation of quantum mechanics is to accept the theory as a complete description of reality, observers included. (In principle, the theory may be applied to the entire universe.)

If we adopt this point of view, we can interpret the simple experiment described above to mean that *both* alternative worlds are equally real, so that when the electron hits the target the universe divides into two copies: one with a left-moving electron, the other with a right-moving electron. A better way of thinking about it is that, before the collision, there are two identical copies of the universe, which then differentiate from each other at the moment of collision. Any observers watching the show must also divide into two copies, one who sees the electron going one way, another who sees it going the other way. Either observer may be fooled into believing that theirs is the only 'real' world, the other being a potential but unrealized contender. But in fact all possible quantum realities co-exist in parallel. This set of ideas became known as the 'parallel universes' or 'many universes' interpretation of quantum mechanics. In general, subatomic activity will create not just two, but a countless number of parallel universes, a process which will be going on all the time.

In spite of its mind-boggling nature, the many universes interpretation of quantum mechanics is probably the most popular among physicists working on fundamental topics such as string/M theory. It is also the favoured interpretation when quantum mechanics is applied to cosmology.

exact trajectory of the particle, and quantum uncertainty tells us that there is a probability that the particle might fail to rearrange the letters, or make a different rearrangement. Deutsch invites us to consider this scenario in the context of the quantum multiverse – the many universes interpretation of quantum mechanics (see Box 8, above). If we could observe all of the parallel quantum realities simultaneously (which isn't possible for human beings, but consider it as a thought experiment), then we would notice that the DNA sequences of a given organism (for example, a particular dog) in these parallel worlds differed. That is because the different quantum worlds have different histories, including the evolutionary histories of the said organism – histories that may depend critically on cosmic ray impacts and other atomic goings-on that are subject to quantum uncertainty. On closer inspection, we would find that the changes in the DNA sequences as we scanned across the many parallel worlds are overwhelmingly concentrated in the junk part of the DNA, while the coding, or genetic, parts would be more or less the same from one quantum world to another, having been conserved by natural selection operating in the various past histories. Deutsch makes the point that, by extending our world view to encompass the quantum multiverse, we perceive a real, physical difference between the genetic sequences, which have 'knowledge', and the junk sequences. In this manner, 'knowledge' may be given a fundamental and objective physical basis.

The upshot of these various arguments is that living organisms are too special in many important ways to be 'just another sort of physical system' to be added to rain clouds and rocks and glue in the vast inventory of nature's bric-a-brac. Viewed cosmologically, the history of the universe is one in which fundamentally new phenomena emerge at successive thresholds of temperature, energy or complexity. For example, at about one microsecond, quarks and gluons congealed into protons and neutrons. At 380,000 years, electrons and nuclear particles combined into atoms. After a few hundred million years, galaxies and stars formed. Some time later, life emerged, then mind, then culture. Nobody would deny that atoms, stars and galaxies are fundamental features of the universe. It seems clear that life (and, as I shall argue in what follows, mind and culture too) is an equally significant step on the path of cosmic evolution.

Taking mind seriously

When it comes to the mental realm, the characteristic qualities are even more distinctive and totally unlike anything else found in nature. Now we are dealing with thoughts, purposes, feelings, beliefs – the inner, subjective world of the observer, who experiences external reality through the senses. These mental entities are clearly not merely 'other sorts of things' – they are in a class apart. They do not even exist on the same level of description as material objects, and bear no obvious relationship to them whatsoever. Open up a brain and you don't see thoughts and feelings, just complicated arrangements of matter. To be sure, it is possible to determine the correlations between neural states and mental states (what philosophers have come to call 'the easy problem', though it is far from easy), but this still leaves untouched the problem of how the subjective experience of, say, the redness of red differs from the blueness of blue or the taste of salt or the feeling of fur. Explaining these so-called qualia has been dubbed part of 'the hard problem'[12] because it is conceptually as well as scientifically disconnected from the world of material objects and forces. Either qualia are illusory, and can be defined out of existence (you don't really see red, you just fool yourself; you don't really exist, you just hallucinate your sense of self), or they are truly fundamental, emergent properties of nature. Heroic attempts have been made to argue the former position, most famously by Daniel Dennett,[13] but the issue is very far from resolved, in my opinion.

Because the literature on this topic is so extensive, I shall not attempt to summarize here the robust defence that has been offered for the fundamental nature of mental states in general, and qualia in particular.[14] I shall just give two additional reasons why I think mind must be taken seriously as a deep and meaningful feature of the universe; the first scientific, the second philosophical. It is true that the mental realm is currently deeply mysterious, but I am convinced that the phenomenon of consciousness will eventually be integrated into the scientific picture of the world, and that the relationship between mind and matter will be properly understood without having to resort to defining consciousness out of existence. Just how this will be achieved remains, of course, a matter of conjecture, but there is at least a hint of how mind might fit into physics that comes from the subject of quantum mechanics. Although quantum systems are intrinsically uncertain, performing a measurement will normally result in a definite outcome (see Boxes 4 and 8, pp. 72 and 257). Since the inception of quantum mechanics, the key role of the act of measurement, or observation, has been recognized, and while it remains unclear how, or whether, mind as such (as opposed to brains or other complex information-gathering systems) enters into this issue, it seems likely that any attempt to bring consciousness within the scope of physics will need to be formulated within the context of quantum mechanics.

The problem of including the observer in our description of physical reality arises most insistently when it comes to the subject of quantum cosmology – the application of quantum mechanics to the universe as a whole – because by definition 'the universe' must include any observers. Andrei Linde has given a deep reason for why observers enter into quantum cosmology in a fundamental way. It has to do with the nature of time. The passage of time is not absolute; it always involves a change of one physical system relative to another, for example, how many times the hands of a clock go round relative to the rotation of the Earth. When it comes to the universe as a whole, time loses its meaning, for there is nothing else relative to which the universe may be said to change. This 'vanishing' of time for the entire universe becomes very explicit in quantum cosmology, where the time variable simply drops out of the quantum description.[15] It may readily be restored in the theory by considering the universe to be separated into two subsystems: an observer with a clock, and the rest. Then the observer may measure the passage of time relative to the evolution of the rest of the universe. So the observer plays an absolutely crucial role in this respect. Linde expresses it graphically: 'Thus we see that without introducing an observer, we have a dead universe, which does not evolve in time,'[16] and 'We are together, the universe and us. The moment you say that the universe exists without any observers, I cannot make any sense out of that. I cannot imagine a consistent theory of everything that ignores consciousness . . . In the absence of observers, our universe is dead.'[17] And observers will exist, obviously, only in those 'Goldilocks' universes in which the laws and conditions are such as to permit them to emerge.

Let me now turn to the philosophical argument for why I believe that mind occupies a significant place in the universe. It concerns the fact that minds (human minds, at least) are much more than mere observers. We do more than just watch the show that nature stages. Human beings have come to *understand* the world, at least in part, through the processes of reasoning and science. In particular, we have developed mathematics, and by so doing have unravelled some – maybe soon, all – of the hidden cosmic code, the subtle tune to which nature dances. Nothing in the entire multiverse/anthropic argument (and certainly nothing in the unique, no-free-parameters theory) requires *that* level of involvement, *that* degree of connection. In order to explain a bio-friendly universe, the selection process that features in the weak anthropic principle merely requires observers to observe. It is not necessary for observers to understand. Yet humans do. Why?

I am convinced that human understanding of nature through

science, rational reasoning and mathematics points to a much deeper connection between life, mind and cosmos than emerges from the crude lottery of multiverse cosmology combined with the weak anthropic principle. In some manner that I shall endeavour to explicate shortly, life, mind and physical law are part of a common scheme, mutually supporting. Somehow, the universe has engineered its own self-awareness. I shall argue in the coming sections that the bio-friendliness of the universe *is* an observer selection effect, but that it operates at a much deeper level than the passive 'winners in a random lottery' explanation.

If it is true that life and mind are fundamental properties of the universe, as I have proposed, then we might expect them to be wide-spread in the universe, just as other fundamental entities (galaxies, stars, atoms) are widespread. By contrast, if life on Earth is just some sort of quirky 'chemical scum', resulting not from any deep principles but just from improbable accidents of fate, then it is likely to be restricted to the solar system. There are then two ways to test the claim that life is a fundamental, significant and universal phenomenon. The first is to search for other instances of life elsewhere in the universe that have arisen from scratch, independently of life on Earth. The second is to look for evidence that life on Earth has started more than once: perhaps there is an alternative form of microbial life still thriving somewhere within Earth's biosphere.[18] If either of these discoveries were made, it would suggest that life is somehow built into the deep nature of the universe – that it is a 'cosmic imperative', to repeat de Duve's evocative description.

The claim that life and observers are the inevitable outcome of an intrinsically bio-friendly universe undoubtedly has great appeal. However, it does run into some serious scientific and philosophical obstacles. To see why, consider a couple of examples of physical systems unarguably predestined to emerge as the inevitable end prod-ucts of physical laws. The first is the crystal. A crystal's structure is determined by the geometrical symmetries built into the laws of electromagnetism. The process whereby an amorphous solution of, say, salt crystallizes into a solid salt grain with a specific geometrical structure is well defined and deterministic: the same thing will happen every time. The second example is the thermodynamic equilibrium of a gas.[19] If a gas is introduced into a closed container in an arbitrary state and left to itself, it will rapidly approach a final state in which the temperature and pressure become constant throughout the gas, and the velocities of the molecules are distributed according to a precise mathematical relationship (known as a Maxwell–Boltzmann distribution). Again, the final state is entirely predictable and repeatable.[20] It is determined in advance by the forms of the laws of physics. It is thus entirely correct to claim that the end states of the salt and the gas are 'written into' the laws of physics.

The question before us, however, is whether life, and maybe even consciousness, is written into the laws of physics. Could the emergence of life from non-life resemble crystallization, say, and follow, predictably and inevitably, from the laws of physics alone, under a wide range of initial conditions? The answer is a decisive no. Biological systems fall between the twin extremes of a crystal and a chaotic gas. A living cell is distinguished by its immense organized complexity. It has none of the simplicity of a crystal and none of the disorder of the gas.[21] It is a specific and peculiar state of matter with high information content. The genome of the smallest known bacterium contains millions of bits of information – information which is not encoded in the laws of physics. The laws of physics are simple mathematical relationships expressible with very little information. They are universal laws, and because they apply to everything they cannot contain information specific to one class of physical system – that is, to living organisms. To understand the high information content of life we must recognize that it is a product, not of the laws of physics alone, but of the laws of physics *and* the history of the environment together. Life emerged and evolved its immense complexity as the result of a process which took billions of years and required a vast number of information-processing steps. A biological organism therefore encapsulates the products of a complex and convoluted history. To sum it up in a phrase, life as we observe it today is 1 per cent physics and 99 per cent history.

Tackling the t-word

If life is not written into the laws of physics as we currently know them, is it possible that those laws can be augmented by some organizing principle which facilitates the emergence of biological complexity, fast-tracking matter and energy along the road to life against the raw odds, and driving it to ever more complex forms? Such a principle has been suggested many times,[22] but always in the face of fierce opposition from orthodox science. And the reason for the negative reaction is not hard to identify. Any sort of life principle or cosmic imperative reintroduces into science the dreaded t-word: *teleology*. The word derives from the Greek *telos* meaning 'end' or 'outcome', and it goes to the heart of what scientists mean when they use the word 'cause'.

Aristotle taught that causes come in different varieties, one of which is what he termed the 'final cause': the end state towards which actions are directed. Final causes are familiar in human activities. For example, a builder goes to buy bricks in order to build a house, a cook puts food in the oven in order to make a meal. The concepts of the house and the meal as end states form part of the chain of causation. We could not fully understand what the builder or the cook are doing without taking into account this 'teleological dimension' of their activities, what we may loosely call their purpose. We can also glimpse teleology at work in the actions of other animals: the dog digs in the garden in order to unearth the bone, the hawk swoops in the hope of catching the mouse. Aristotle believed that he could discern final causes in inanimate nature too, and we still see an echo of this thinking in phrases such as 'water seeks its own level'. But since the time of Newton, final causation in physics has been banished (or at least strongly de-emphasized). At the level of atoms interacting with other atoms, causation is right there in the interaction, and makes no reference to any sort of destiny or purpose or final outcome.

When it comes to biology, teleology in behaviour is hard to deny – certainly in human behaviour. But teleology was decisively eliminated from biological *evolution* by Darwin. The core idea of Darwinism is that nature cannot 'look ahead' and anticipate what may be needed for survival. To use Richard Dawkins' phrase, nature is indeed a blind watchmaker: mutations occur randomly and are selected for their survival value *at the time*. Darwinism acts in the here and now. Selection selects for the fittest at any given moment. According to Darwin, evolution isn't 'going anywhere': there is no directionality, no advance planning.[23] To be sure, evolution may display trends, such as the progressive growth in the elephant's trunk, but that is simply because of a relatively constant selection pressure amplifying a useful trait. In this respect Darwinism stands in sharp contrast to the (now discredited) alternative theory of evolution proposed by Lamarck, the inheritance of acquired characteristics, which I mentioned in passing in Chapter 9. In Lamarck's theory, organisms strive to achieve better adaptation and pass on the fruits of their efforts to their offspring: for example, the giraffe stretches out its neck to reach higher branches, and as a result produces longer-necked baby giraffes. So Lamarckianism contains a clear teleological component. The triumph of Darwinism, however, has given any sort of teleology, whether in biology, physics or elsewhere, a very bad press.[24] I would even go so far as to say that science is in the final throes of teleological cleansing.[25] The antipathy stems partly from the theological overtones associated with teleology. Although Aristotle's original concept of final causation was theologically neutral, teleology came to be seen by scientists as tantamount to the guiding hand of God at work in the physical universe. That was anathema, and the overthrow of teleology in evolution was greeted with enthusiasm by atheists. Friedrich Engels, for example, had this to say in a letter to Karl Marx in 1859: 'Darwin, by the way, whom I'm reading just now, is absolutely splendid. There was one aspect of teleology that had yet to be demolished, and that has now been done.'[26]

The strong anthropic principle and even de Duve's at first sight innocuous 'cosmic imperative' flirt with teleology. They describe the facilitation of a particular end state of affairs – life, consciousness – via a long sequence of steps, a sequence which culminates billions of years after the laws of nature have been 'laid down'. Hard-headed scientists pour scorn on such starry-eyed notions. Gell-Mann probably speaks for the majority when he writes that, 'Life can perfectly well emerge from the laws of physics plus accidents, and mind from neurobiology. It is not necessary to assume additional mechanisms or hidden causes,'[27] and 'In its strongest form, however, such a [teleological] principle would supposedly apply to the dynamics of the elementary particles and the initial conditions of the universe, somehow shaping those fundamental laws so as to produce human beings. That idea seems to me so ridiculous as to merit no further discussion.'[28]

Teleology is unpopular with scientists not only for ideological reasons: there are sound scientific arguments against it too. According to the conventional view, the laws of physics already determine everything that happens in nature (to within quantum uncertainty), once the initial conditions of the system of interest have been specified. If we now attempt to superimpose an extra law or principle on top of the laws of physics, it looks like there will be a conflict. For example, if the laws of physics tell an atom to do this, but the teleological principle says it should do that, how is the poor atom to decide? Physicists call this a case of over-determinism: the system is already 'causally saturated' at the micro-level by the basic laws of physics, and there is 'no room at the bottom' for any competing imperative.

Abandoning Platonism would make room for teleology

There is no possibility of introducing a strong anthropic principle or a biological imperative into cosmology as long as the origin and evolution of the universe are already determined by the laws of physics as we at present conceive them (e.g. by string/M theory). But this seemingly unassailable conclusion conceals a weakness, albeit a subtle one. The objection that there is no room at the bottom for an additional principle rests on a specific assumption about the nature of physical laws: the assumption called Platonism, which I mentioned briefly in Chapter 1. Most theoretical physicists are Platonists in the way they conceptualize the laws of physics, as precise mathematical relationships possessing a real, independent existence which nevertheless transcends the physical universe (I depicted this point of view in Figure 2, p. 15). For example, in simple, pre-multiverse cosmological models, where a single universe emerges from 'nothing' (see p. 87), the laws of physics are envisaged as 'inhabiting' the 'nothingness' that preceded space and time. Heinz Pagels expressed this idea vividly: 'It would seem that even the void [the state of no space and no time before the big bang] is subject to law, a logic that exists prior to time and space.'[29] Likewise, string/M theory is regarded as 'really existing, out there' in some transcendent Platonic realm. The string theory landscape, with its convoluted and elaborate shape determined by specific equations of physics, also exists 'out there'. The universe-generating mechanism of eternal inflation exists 'out there'. Quantum mechanics exists 'out there'. Platonists take such things to be independently real – independent of us, independent of the universe, independent of the multiverse. But what happens if we relinquish this idealized Platonic view of the laws of physics?

Anton Zeilinger, an Austrian physicist who works on tests and applications of quantum mechanics, is one who has some reservations: 'The laws we discover about Nature do not already exist as "Laws of Nature" in the outside world.'[30] Many physicists who do not concern themselves with philosophical issues prefer to think of the laws of physics more pragmatically as regularities found in nature, and not as transcendent immutable truths with the power to dictate the flow of events. Perhaps the most committed anti-Platonist was Wheeler. 'Mutability' was his byword. He liked to quip that, 'There is no law except the law that there is no law.'[31] Adopting the catchy aphorism 'Law without law' to describe this contrarian position, Wheeler maintained that the laws of physics did not exist a priori, but emerged from the chaos of the quantum big bang – coming out of 'higgledy-piggledy' was the way he quaintly expressed it – congealing along with the universe that they govern in the aftermath of its shadowy birth.[32] 'So far as we can see today,' he maintained, 'the laws of physics cannot have existed from everlasting to everlasting. They must have come into being at the big bang.'[33] Crucially, Wheeler did not suppose that the laws just popped up, ready-made, in their final form, but emerged in approximate form and sharpened up over time: 'The laws must have come into being. Therefore they could not have been always a hundred per cent accurate.'[34]

The idea that the laws of physics are not infinitely precise mathematical relationships, but come with a sort of inbuilt looseness that reduces over time, was motivated by a belief that physical existence is what Wheeler called 'an information-theoretic entity'. He pointed out that everything we discover about the world ultimately boils down to bits of information.[35] For him, the physical universe was fundamentally informational, and matter was a derived phenomenon (the reverse of the orthodox arrangement), via a transformation he called 'it from bit', where the 'it' is a physical object such as an electron, and the 'bit' is a unit of information.[36]

Why should 'it from bit' imply 'law without law'? Rolf Landauer, a physicist at IBM who helped to lay the foundations for the modern theory of computation, was able to clarify the connection. Landauer also rejected Platonism as an unjustified idealization. What bothered him was that, in the real world, all computation is subject to *physical* limitations.[37] Bits of information don't float freely in the universe: they always attach to physical objects. For example, genetic information resides on the four nucleotide bases that make up your DNA. In a computer, bits of information are stored in a variety of ways, such as in magnetized domains. Clearly, one can't have software without hardware to support it. Landauer set out to investigate the ultimate limits to the performance of a computer, the hardware of which is subject to the laws of physics and the finite resources of the universe. He concluded that idealized, perfect mathematical laws are a complete fiction as far as the real world of computation goes.

To get the drift of Landauer's reasoning, consider the mathematical operations associated with applying Newton's laws. Newton invented the mathematics he needed to describe these laws (Leibniz did too, leading to a debate over priority). Newton called it the theory of fluxions; today we call it calculus. For readers who never learned calculus, the details are unimportant; suffice it to say that it requires certain variables, such as the position of a body and the time at which that body is observed, to vary *continuously*. For example, the speed of a body is the rate of change of position with time. The acceleration is the rate of change of speed with time. To carry out the basic mathematical operations expressed in Newton's laws, you have to assume that space and time are continuous and infinitely subdivisible. The upshot is that Newton's laws, considered as exact mathematical statements, require intervals of space and time to be continuous and smooth on any scale of magnification, right down to zero.

These properties can be illustrated by what mathematicians call 'the real line' – a continuous line on which each point can be labelled by a real number. A real number is a decimal number, such as 0.563 715 738 . . ., almost all of which possess an infinite number of digits after the decimal point. There are no gaps on the real line: the real numbers can be packed arbitrarily closely together. Real numbers play an absolutely fundamental role in physical theory, not only in Newtonian mechanics but in almost all of physics. In practice, of course, we can never observe space and time on arbitrarily small scales. About the best we can do at the moment is to indirectly probe space down to about 10^{-18} cm and time down to about 10^{-28} s. But the mathematical operations involved in solving the equations of Newton's laws are easier if we assume the real-number idealization. That's all well and good when human beings are doing the calculations, but computers cannot handle infinite and infinitesimal quantities: they deal instead in discrete bits, or jumps (in 1's and 0's). The size of the jumps can be made very small in a given calculation, but the smaller they are, the more computing power is required. The jumps are always finite. It would take an infinite amount of computing power to model the real line.

The question Landauer asked is whether the mathematical idealizations embodied in Newton's laws and the other laws of physics should really be taken seriously. As long as the laws are confined to some abstract realm of ideal mathematical forms, there is no problem. But if the laws are considered to inhabit, not a transcendent Platonic realm but *the real universe*, then it's a very different story. The real universe will be subject to real restrictions. In particular, it may have finite resources: it may, for example, be able to hold only a finite number of bits at one time. If so, there will be a natural *cosmic* limit to the computational prowess of the universe, even in principle. Real numbers, on which the orthodox notion of most physical laws is based, could not exist.

Gregory Chaitin, who, like Landauer before him, works for IBM and is a leading theorist on the conceptual foundations of computation, has arrived at the same conclusion. He expresses it graphically: 'Why should I believe in a real number if I can't calculate it, if I can't prove what its bits are, and if I can't even refer to it? . . . The real line from 0 to 1 looks more and more like a Swiss cheese.'[38] Landauer's point of view was that there is no justification for invoking mathematical operations to describe physical laws if those operations cannot actually be carried out, even in principle, in the real universe, subject as it is to various physical limitations. In other words:

Laws of physics that appeal to physically impossible operations must be rejected as inapplicable.

Platonic laws can perhaps be treated as useful approximations, but they are not 'reality'. Their infinite precision is an idealization that is normally harmless enough, but not always. Sometimes it will lead us astray, and never more so than in discussion of the very early universe.

The universe as a finite computer exposes the fiction of idealized laws

To see where the problem lies, let us estimate how the real universe, with its finite resources and processing power, measures up to the Platonic ideal. As we saw in Chapter 3, the observable universe is finite because the finite speed of light implies the existence of a horizon. Because no physical object or influence can go faster than light, objects separated by more than the distance to the horizon cannot communicate with each other. So Landauer's criterion says that the great cosmic computer we call the observable universe must be limited to objects encompassed by a volume of space that is less than the distance to the horizon – the region I have been calling the observable universe. At the present epoch, the volume of space within the horizon contains about 10^{80} atoms, and about 10^{90} neutrinos and photons. Each particle can carry a few bits of information only. Additional information can be encoded in gravitons, which cosmologists believe permeate the universe, although nobody seriously expects to detect any in the foreseeable future. A careful calculation has been carried out by Seth Lloyd, a theoretical physicist at the Massachusetts Institute of Technology, and he comes up with a figure of about 10^{120} bits in total.[39]

The actual number is less important than the fact that the total amount of information contained in the universe, though admittedly huge, is nevertheless finite.

According to Landauer's philosophy, it is pointless applying any law of physics at a level of detail which requires the processing of more bits of information than the cosmic upper limit of 10^{120}, because there is an intrinsic inaccuracy (or 'higgledy-piggledy' in Wheeler-speak) which is quantified by this huge number. To take a specific example, the law of conservation of electric charge states that the charge on an electron should be *exactly* constant with time. According to Landauer's view, this statement is meaningless because it implies infinite precision. Instead, one should imagine that the law applies only with a finite accuracy of one part in about 10^{120}. Since we can currently measure the electron's charge to an accuracy of only about one part in 10^{12}, this is hardly a serious restriction. For almost all day-to-day purposes it doesn't matter whether the universe is considered to be a finite computer with limited accuracy, or a system conforming to infinitely precise mathematical laws.[40]

Although the sloppiness of the laws of physics implied by the cosmic upper limit derived by Lloyd is largely unimportant today, it may have been very important in the past. That is because the radius of the horizon isn't fixed, but increases with time at the speed of light. The number of particles contained within a volume of space bounded by the horizon is therefore going up year by year as the horizon expands to encompass more and more matter – so in the past, this number was smaller. At one second after the big bang, for instance, the horizon encompassed only about 10^{86} particles – still too large for the implied inaccuracy to make much difference. At the time of inflation, however, the horizon was a mere trillion-trillionth of a centimetre in radius, and the total information content of a horizon volume was then only about a billion bits. Such a small number of bits represents a very large degree of looseness, or ambiguity, in the operation of any physical laws, including the laws of string/M theory (or whatever theory is supposed to govern the inflationary process). I mentioned Wheeler's suggestion that the laws of physics emerged from 'higgledy-piggledy' at the big bang in a less than precise form, and gradually 'congealed' over time. In this section I have shown how, by accepting that the universe is a finite computational resource, and making use of the work by Landauer and Lloyd, Wheeler's suggestion can be made explicit.

Mathematics and physics emerge as one

Can the ideas set out in the previous section help us to understand the deep mystery of why the universe is mathematical? In Chapter 1, I wrote of the 'hidden subtext' of nature, the fact that the laws that govern the physical world are mathematical in form, and how by stumbling across this 'cosmic code', scientists have been able to unravel what makes the universe tick. In orthodox physics, the fact that nature conforms so efficiently to elegant mathematical principles is left completely unexplained. Physicists are obliged to assume the existence of two separate realms: the Platonic world of perfect mathematical objects and relationships lying outside the physical universe, and the world of space, time and physical objects. It is then taken for granted that a deep link exists between these two realms. Paul Benioff of the Argonne National Laboratory, one of the pioneers of quantum computation, expresses this assumption graphically: 'In many ways theoretical physics treats mathematics much like a warehouse . . . If a system needed by physics has been studied, it is taken from the warehouse, existing theorems are used, and, if needed, new theorems are proved.'[41] Benioff argues that a final theory of everything should not merely unify all of physics, but should also provide a common explanation for physics and mathematics. In other words, mathematics and physics should not be assumed to exist separately a priori, but should emerge together from a single coherent theory of existence. In this manner, the efficacy of mathematics in describing the physical world would be automatically incorporated into the unification scheme.

As long as scientists are fixated on immutable Platonic laws which transcend the physical universe, the unification of physics and mathematics will be impossible, and the mathematical nature of physical laws will remain mysterious. But taking the informational view described in the foregoing sections, in which mathematics is tied to the physical world even as the physical world is tied to mathematical laws, offers the hope of complete unification. To achieve this goal, it will be necessary to describe a self-consistent intertwining of mathematics and physics.[42] An interesting unsolved question is then whether the criterion of self-consistency is stringent enough to single out certain laws: for example, will quantum mechanics be required as the fundamental theory of matter?

Part of the reason why mathematicians are ambivalent about the nature of mathematics is that on the one hand doing mathematics seems like a voyage of discovery – mathematicians 'come across' mathematical objects and relationships which in some sense already exist – but on the other hand mathematics is a product of the human intellect. So is mathematics discovered, or invented? Platonists believe the former – that mathematics has an independent reality, and that humans have been fortunate to access it and find such fruitful applications to nature. But if Platonism is rejected, then we must accept that mind or intellect plays a much deeper role in the explanation of the physical universe, as advocates of the strong anthropic principle maintain. Benioff arrives at a similar conclusion. The union of mathematics and physics comes, he argues, from the existence of a common underlying theory. 'Since intelligent beings are necessary to create such a theory,' he says, 'it follows that the basic properties of the physical universe must be such as to make it possible for intelligent beings to exist ... None of this implies that intelligent beings must exist, only that it must be possible for them to exist.'[43] In this manner, mathematics, physics, life and mind would emerge from a self-consistent, common axiomatic scheme.

The main argument against the existence of any sort of universal principle favouring the emergence of life and mind – for example de Duve's cosmic imperative, or Carter's strong anthropic principle – is that the basic laws of physics plus initial conditions already fix what physical systems do, and there is simply no more room in which an additional teleological law can operate. But if the basic laws of physics are not in fact rigid in the Platonic sense, if there is a looseness or inherent limitation on the accuracy of those laws – especially in the early moments of the universe, when its bio-friendly nature was being laid down – then a loophole exists for a law-like trend towards life and mind to peacefully co-exist alongside the traditional laws of physics. There would no longer be any conflict.

Permitting a trend towards life is one thing; realizing it is another. Teleology has been out of favour not only because of its perceived conflict with the laws of physics. It also suffers from a seemingly insurmountable problem to do with cause and effect. Teleology is by definition a means to anticipate some future state (in this case life) and bring that state about in the fullness of time. This blatant element of predestination is in sharp contrast with the normal concept of causation in science, in which present events can influence the future but not the past. Teleology turns that around, and lets future states influence the present. How can that be? How could the very early universe – the epoch when the laws of physics were still in the melting pot – possibly *know* about life and mind emerging billions of years later?

Quantum mechanics could permit a subtle form of teleology

Crazy though the idea may seem at first, there is in fact no fundamental impediment to a mechanism that allows later events to influence earlier events. In fact, there are some famous theories of physics that explicitly involve *backward causation* – future events having causative power over past events. Wheeler proposed one such theory with his then student Richard Feynman in the mid-1940s.[44] In the Wheeler–Feynman theory of electrodynamics, electromagnetic interactions can travel both forwards and backwards in time. There is no experimental evidence in favour of the theory, I hasten to add. Something similar was proposed for gravitation by Hoyle and Narlikar,[45] and for quantum cosmology by Gell-Mann and Hartle,[46] and by Hawking.[47] Again, observation and experiment are silent on these ideas, but the theories are certainly not 'anti-scientific', and variants of them are still being investigated today. It is only when the end states involve life and mind that most scientists take fright and bale out. That is because life and mind are not normally regarded by physicists as fundamental. Furthermore, they are tainted with mystical associations from a bygone era of vital forces. But, as I have argued earlier in this chapter, a good case can be made that life and mind *are* fundamental physical phenomena, and so must be incorporated into the overall cosmic scheme. One possible line of evidence for the central role of mind comes from the way in which the act of observation enters into quantum mechanics. It turns out that the observation process conceals a subtle form of teleology. To see why, we need to examine the details of a particular quantum measurement.

A well-known manifestation of quantum weirdness is the fact that a quantum entity such as a photon can behave sometimes like a wave and sometimes like a particle (see Box 4, p. 72). Which aspect of the photon is manifested depends on the experimental arrangement used to observe it. For example, when a photon hits a photographic plate and makes a little spot, it reveals its particle nature. But one can make photons do wave-like things too, by using different equipment. Consider, for example, a famous experiment first performed in the eighteenth century by the English physicist and Egyptologist Thomas Young. The set-up is shown in Figure 31. It consists of a pinpoint source of light and a screen with two slits cut in it to let the light through. The image of the slits is inspected on a second screen. You might think that the image would consist of two closely spaced and overlapping fuzzy patches of light. In fact, it is a series of bright and dark lines, called interference fringes, which arise because of the wave nature of light. The waves pass through both slits, and spread out on the far side.[48] Where the wave from one slit merges with the wave from the other slit, the two sources of light will combine. If the waves arrive in step with each other (i.e. with the same phase), they will reinforce; where they are out of step they will cancel. Young's bright and dark interference fringes provided the first conclusive evidence that light is a wave.[49]

Things become baffling, however, when we take account of the particle nature of light – the photon aspect. In normal experience a particle is localized in space, whereas a wave is spread out. A wave can go through both slits and recombine – this is the basis of the interference phenomenon – but a particle must surely pass through *either* one slit *or* the other. If light behaves as a stream of particles, like bullets from a machine gun, then there can be no interference

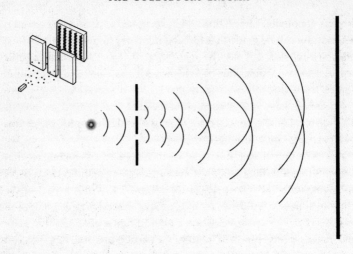

31. *Wave or particle – or both?*

Young's famous two-slit experiment is shown viewed from above. Light from a point source falling on a screen which has two apertures makes an image on a second screen. The image appears in the form of bright and dark vertical bands called interference fringes (shown in the three-dimensional image at top left), thus demonstrating the wave nature of light. But light may also be considered as being made up of particles (photons). The light source can be dimmed to the point where only one photon at a time passes through the apparatus, each photon arriving at a specific point on the image screen. When many such dots have accumulated, a speckled interference pattern is discerned, so even individual photons must 'know about' both slits, even though it would seem that a given photon will pass through either one slit or the other. If the experimenter deliberately looks to see which slit each photon passes through, the interference pattern will not form, and the wave-like nature of light is lost: it behaves purely as a stream of particles.

because any given photon would go through only one of the two slits, and know nothing of the other one. That much seems obvious. So what happens if the experimenter turns the intensity of the light source down so low that only one photon as a time traverses the apparatus? The arrival of each individual photon can be recorded, say by a photographic plate. The photon creates a little spot on the plate. Each photon arrives at a particular place on the screen – it does not get spread out to make a pattern. But when the experiment is allowed to run, accumulating the results of many photons, then an interference pattern starts to emerge, in a speckled sort of way, from the individual photon events. So although the arrival of the light is recorded as individual particle-like dots, the collective effect is to produce a wave interference pattern. Now this result is weird, because it seems to imply that any given photon must somehow know about both slits so that it can cooperate with the other photons in creating the collective interference pattern – despite the fact that a particle should be able to pass through only one slit. Sometimes this is expressed by saying that the photon went through *both* slits, that it was in two places at once! However – and this is crucial – the interference pattern will emerge only if the experimenter makes no attempt to determine which slit any given photon went through. Any sneaky apparatus stationed near the slits to glimpse the photon taking one path or the other will mess up the experiment. If the experimenter succeeds in detecting a photon going through a given slit, that photon will not contribute to the interference pattern.

The two-slits experiment dramatically illustrates the key role that the experimenter, or observer, plays in determining the nature of quantum reality. But what does this have to do with backward causation? Buried in the subtleties of the discussion so far is an unanswered question: when, exactly, did nature 'decide' to opt for wave or particle? In the 1980s, Wheeler zeroed in on this question by envisaging a refinement of Young's experiment. His idea was to convert the image screen into a venetian blind, and position a pair of detectors (e.g. small telescopes) behind it, each pointing at one of the slits (see Figure 32). If the blind is left closed, the system functions as in the original experiment, with the wave nature of light manifested as an interference pattern. But if the blind is opened, allowing the photons to pass through, the detectors can be used to work out which slit the photon came from: if detector A registers a photon we know that it must have gone through slit 1, whereas if B registers a photon, we know it that it must have gone through slit 2. (If the photon misses either detector we can't say anything.) In this mode of operation the particle nature of light is manifested. The experimenter can decide

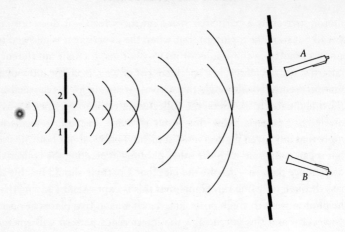

32. *Delayed-choice experiment*

John Wheeler imagined adapting the set-up shown in Figure 31, replacing the image screen by a venetian blind (shown here with the slats running vertically out of the page) and stationing a pair of telescopes behind it, each directed at one of the two slits. When a photon approaches the blind, the experimenter may choose to leave it closed, and recover the results of the conventional Young's experiment (the fringes shown in Figure 31), or open it and permit the telescopes to register which slit the photon passed through. But how does the photon 'know', at the time it passes through the first screen, what the experimenter will decide? Spur-of-the-moment decisions made by the experimenter affect the nature of reality (in this case, wave or particle) as it *was*, in the past.

on a photon-by-photon basis which experimental configuration to employ, and thus which aspect of light shall be manifested – wave, wave, particle, wave, particle, particle, wave, ... in an arbitrary sequence.

Now we get to the punchline. The experimenter can *delay the choice* – wave or particle – right up to the moment the photon arrives at the venetian blind. The mystery we then have to confront is *when* the photon adopted the form – wave or particle – chosen by the experimenter. How could a photon know, in advance of the measurement, whether the blind would be opened by the experimenter or

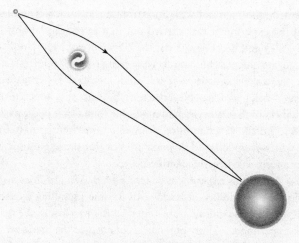

33. *Cosmic delayed-choice experiment*

The experiment described in Figure 32 could in principle be performed in an astronomical context. Light from a distant quasar is bent by the gravitational spacewarp of a galaxy and directed towards Earth. Photons can reach Earth via two paths, mimicking the slits of the original Young's experiment (Figure 31).

not? Does it defer a decision – wave or particle – right up until the experimenter makes the choice? That can't be quite right, because if the photon is a particle it passes through only one slit, whereas if it is a wave it passes through both. It needs to know, when it reaches the screen with the two slits, which to be, wave or particle (i.e. to go through both slits or one). It can't wait until it reaches the venetian blind.[50]

Wheeler envisaged a 'natural' Young's experiment in which the two paths are defined not by slits in a screen, but by the gravitational lens effect of an interposing galaxy bending light from a distant quasar (see Figure 33). If the delayed-choice experiment could at least in principle be extended to intergalactic dimensions in this manner, the photons would need to know what to do (one path or both round the galaxy) *billions of years before Earth even existed*, let alone before the experimenter makes a choice!

Remarkably, Wheeler's delayed-choice experiment has been

performed, although not with an astronomical source. The practical test of the experiment was carried out in a terrestrial laboratory by Caroll Alley and his colleagues at the University of Maryland.[51] Alley replaced the sluggish human decision-maker with an ultra-fast electronic switch flicked at random, and modified other aspects of Wheeler's original idea. Nevertheless, he was able to confirm the essential claim that Wheeler had made. He found that photons which, in effect, 'hit the venetian blind' formed an interference pattern, and those which were allowed through did not (for those photons whose provenance was correctly identified).

How should this experiment be interpreted? What it does *not* do is establish the possibility of backwards-in-time signalling as such: no actual information can be transmitted to the past using the delayed-choice experiment. (Many popular accounts give the incorrect opposite impression.) If an accomplice stationed near the slits tried to find out what the experimenter was going to do by peeking at each photon as it passed by, the very act of observation by the accomplice would compromise the experiment.[52] The best way to think about the delayed-choice experiment is to regard the photon as in some sense less than real in the absence of an observation. I don't wish to give the impression that the photon doesn't exist at earlier times; the point is that, in the absence of an actual observation or measurement process, its state – which can be precisely specified by quantum mechanics – does not define a wave or a particle or even 'a bit of both'. The particle/wave designations come only in the context of an actual experiment. The weirdness of the delayed-choice experiment is that although it is the action of the experimenter that is responsible for fixing the photon's wave or particle nature, the observation that is made has a crucial relevance to the past – maybe even to the very remote past. So what an experimenter may choose to do today helps to shape the nature of reality (e.g. wave or particle) *that was*, perhaps a very long time ago. That is not quite the same as explicit backward causation (which would enable the experimenter to send a signal into the past), but it does have a distinct teleological feel to it. I would describe it, in Wheeleristic terms, as 'teleology without teleology'.[53]

I began this chapter by introducing the so-called strong anthropic principle, a loose set of ideas which seek to establish that the emergence of life and mind in the universe are somehow predestined and inevitable – they are built into the nature of the universe at the deepest level. But as we have seen, in order to implement such a notion two requirements are necessary. First there is the need for an overarching principle, a life principle if you like, to be somehow accommodated by the already existing, life-blind, basic laws of physics. The second is for some element of teleology to creep back into cosmology. I have suggested that the former problem could be solved by abandoning a rigid Platonic view of the nature of physical laws and replacing it with an information-theory picture in which the familiar laws of physics come with an inbuilt level of looseness or flexibility – a level which is minuscule today, but significantly higher in the universe's very early moments when its life-friendly laws and parameters were being established.

The second problem – about some sort of teleology – might be solved by quantum mechanics. Certainly Wheeler believed so. He thought of observers as *participators* in shaping physical reality, and not as mere spectators. In itself that is hardly new: philosophers are steeped in that tradition. The novel feature Wheeler introduced via his delayed-choice experiment was the possibility of observers today, and in the future, shaping the nature of physical reality *in the past*, including the far past when no observers existed. That is indeed a radical idea, for it gives life and mind a type of creative role in physics, making them an indispensable part of the entire cosmological story. Yet life and mind are the *products* of the universe. So there is a logical as well as a temporal loop here. Conventional science assumes a linear logical sequence: cosmos → life → mind. Wheeler suggested closing this chain into a loop: cosmos → life → mind → cosmos. He expressed the essential idea with characteristic economy of prose: 'Physics gives rise to observer-participancy; observer-participancy gives rise to information; information gives rise to physics.'[54] Thus the universe explains observers, and observers explain the universe.[55] Wheeler thereby rejected the notion of the universe as a machine subject to fixed a priori laws, and replaced it with a self-synthesizing world he called 'the participatory universe'.[56] By postulating a closed explanatory loop, similar to the self-consistency argument of Benioff that I considered in the previous section, Wheeler deftly circumvented the infamous tower-of-turtles problem. There is no need for a levitating super-turtle if the bio-friendly universe explains itself.

The universe and mind become one in the far, far future

It is a huge leap from the delayed-choice experiment, which deals with single photons, to the *entire universe* being somehow created (or at least projected into a definite, concrete form) by its own observer-participators. What about all those photons, not to mention other particles, that don't get observed? Remember, however, that observers don't have to be human – they could be any form of sentient being in the universe. More importantly, the observations do not have to happen now. Because of the backwards-in-time aspect of quantum mechanics, the past can be shaped by observations at any stage in the cosmological future.

Humankind has walked this planet for what, in cosmological terms, is but the twinkling of an eye. Earth should remain habitable for at least another billion years. That is plenty of time for our descendants, natural or artificial, flesh-and-blood or machine (or a blend of both), to decamp to another locale. It will be hundreds of billions of years before stars become a rarity. Even then, there will still be black holes – the dead remnants of stars – which store a colossal amount of potentially usable energy. There is no fundamental reason why life and mind could not endure for trillions upon trillions of years. We can certainly imagine, as do many science fiction writers, that over the countless aeons that lie ahead, life and mind will spread out into the cosmos, perhaps from Earth alone, perhaps from many planets. A progressively larger fraction of the universe will be brought under intelligent control. More and more matter will be used to process information and create a rich mental world, perhaps without limit. Many scientists have speculated that, as the time-line stretches towards infinity, so an emerging distributed super-intelligence will become more and more god-like, so that in the final stage the super-mind will merge with the universe: mind and cosmos will be one. It is a vision sometimes referred to as the *final* anthropic principle.[57] As David Deutsch has put it, 'In the final anthropic principle, or if anything like an infinite amount of computation taking place is going to be true – which I think is highly plausible one way or another – then the universe is heading towards something that might be called omniscience.'[58]

If the universe were to become saturated by mind, then it would fulfil the necessary conditions for Wheeler's participatory principle, in which the *entire universe* would be brought within the scope of observer-participancy. The final state of the universe, infused with mind, would have the power to bring into being the pathways of evolution that lead to that same final state. In this way the universe could both create itself and steer itself towards its destiny. 'The coming explosion of life opens the door', declared Wheeler, 'to an all-encompassing role for observer-participancy: to build, in time to come, no minor part of what we call *its* past – *our* past, present and future – but this whole vast world.'[59]

So we now have a third possible scientific answer to the thematic question of this book: why is the universe fit for life? The first is that it is a fluke. The second is that it is the result of observer selection from a multiverse. The third answer, the one I have outlined in this chapter, is that the universe has engineered its own self-awareness, through quantum backward causation or some other physical mechanism yet to be discovered. We have seen how the first two explanations address the issue of the bio-friendliness of the universe, but they necessarily fall short of providing a complete explanation of the biggest of the big questions – namely, why does the universe exist? They cannot answer the ultimate question of existence because they both require an unexplained starting point – what I have whimsically termed a levitating super-turtle – to be accepted as simply given. But what about the third way? Can the idea of the self-synthesizing universe go beyond the issue of why the universe is bio-friendly and illuminate the question of why the universe exists at all?

Loops in time

On 9 January 2001, John Wheeler suffered a heart attack. 'That's a signal,' he said. 'I only have a limited amount of time left, so I'll concentrate on one question: How come existence?'[60] Unfortunately his deliberations on the participatory universe and the final anthropic principle were not so much a well-formulated idea as what he liked to call 'an idea for an idea'. How can we take his idea for an idea and discover what is entailed by a self-explanatory universe, a universe that contains within itself an explanation of its own existence? We get a pointer from the concept of *causal loops*, which are familiar in time-travel science fiction stories. In a typical story, such as the movie *Back to the Future*, or the *Doctor Who* TV series, the time traveller visits the past and changes something. For example, in the movie the leading character, Marty McFly, visits his mother as a young woman and becomes embroiled in her love life, thus threatening her marriage to his father, and thereby his own subsequent existence. A more brutal account of the same idea is when the time traveller seeks out and murders his mother before she could give birth.

The fascination with such stories is that they seem to lead inexorably to paradox. The past determines the present, so if the past is changed then so is the present, including events affecting the time traveller. But paradox is not inevitable if a self-consistent narrative is established. Consider a variant on the matricide scenario in which a time traveller goes back fifty years and encounters a young girl on the point of being shot dead by a robber. The time traveller intercedes and saves the girl's life. The girl grows up to become the time traveller's mother. This self-consistent causal loop involves circular explanation: the girl's survival and motherhood are explained by the time traveller, and the time traveller is explained by the girl's motherhood. A more striking illustration is where a professor visits the future and reads about a new theorem in a current mathematical journal. He then returns to his original time, tells a student the theorem, and the student subsequently publishes it in a journal – the very journal in which the professor found the theorem! A causal loop is again apparent: the professor's knowledge of the theorem came from the student, and the student's knowledge of the theorem came from the professor. Weird though such causal loops may be, provided self-consistency is respected, no genuine paradox is involved.[61]

Time travel and causal loops are not merely the stuff of science fiction. The theory of relativity, which permits time to be warped by motion and gravitation, predicts circumstances in which physical objects, including observers, can loop back into the past. An explicit model of a universe which permits time travel was found by Kurt Gödel in 1948 using Einstein's general theory of relativity, although it is based on the unrealistic proposition that the universe as a whole is rotating. A better model of a time machine features a so-called wormhole – a type of stargate, or short cut, between distant points in space. In a wormhole time machine, an observer who passes through the wormhole in one direction leaps into the future, and in the other direction into the past.[62]

More dramatic than an object or person making a limited visit to the past is when an object *becomes* its own past self, for example, a particle that loops back in time to an era before which it did not exist. It then stays put (i.e. it doesn't get rapidly transported 'back to the future', but just waits for the future to come, like everything else).[63] In a speculative paper entitled 'Can the universe create itself?', cosmologists Richard Gott III and Li-Xin Li applied this idea to the universe as a whole by adapting the theory of baby universes in a mind-bending way.[64] In their theory, one of the babies 'grows up', then loops back in time to become the mother universe. Actually, I published a cosmic causal loop theory myself in 1972.[65] I considered the possibility that the universe will re-contract to an apparent big crunch, which in fact turns out to be a big bounce. The ultra-dense state near the bounce wipes out all material structures and scrambles all information from the preceding history (technically, it is a state of maximum entropy). The novel feature of my model was that in the subsequent phase of expansion and re-contraction, time's arrow is reversed (entropy falls) relative to the initial cycle of expansion and contraction. At the end of cycle 2, the universe is back to where it started – in the state it possessed at the beginning of cycle 1. It is then

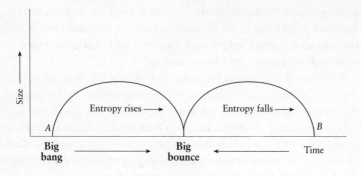

34. *Causal loop universe*

The graph depicts a universe which expands from a big bang *A*, contracts again to a big bounce, and then undergoes a second cycle of expansion and contraction, but with physical processes therein running backwards, as indicated by the long direction-of-time arrows beneath the graph. When the second cycle reaches a big crunch *B*, the universe has returned to its initial state *A*. Time may therefore be closed into a loop, by identifying *B* with A.

a simple matter to identify these two times and close the history of the universe into a loop, like a cosmic version of the movie *Groundhog Day* (see Figure 34).

Causal loops are a good start, but they leave a lot unexplained, for example the laws of physics, so they fall short of providing a complete answer to the question 'How come existence?' One can still ask, 'Why *that* loop?' In the parable of the time-travelling professor who returns from the future with a new mathematical theorem, the question of 'Why that theorem?' is unresolved. To be sure, the theorem comes into existence without a creator, so in that sense we might say that it is 'self-creating'. But the theorem is not *explained* by this device. We can imagine a limitless number of causal loops containing a limitless number of different theorems. In the context of cosmology, one self-creating universe can be accompanied by an infinite number of different self-creating universes – there could be a multiverse of causally closed universes, each with different laws of physics. So we still run into the problem of 'the rule' (see p. 241) – what is it that determines which causal-loop universes have 'fire breathed into them' and really

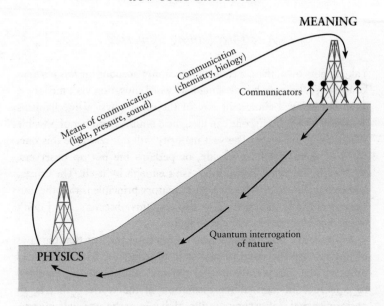

35. *The closed loop of existence: a participatory universe*
Wheeler's 'meaning circuit' depicts physical reality as self-explanatory, or
'self-synthesizing'. Physics gives rise to objects, organisms and, eventually,
communicators of meaningful information about nature. In the return part
of the cycle, observer-participants interrogate nature for bits of information
(ultimately via quantum mechanics), and thus help to shape physical reality,
even in the far past. Observers give rise to physics even as physics gives rise
to observers. In this manner, Wheeler seeks to avoid the tower-of-turtles
problem by claiming that the physical world and its observer-participants
explain each other.

exist, and which languish as potential-but-not-actually-existing universes? (Unless, of course, *all* self-consistent causal-loop universes
actually exist.) A fully satisfactory *explanatory* loop would have to
yield a complete explanation for everything, laws included. It should
also tell us why those laws are bio-friendly. That was precisely what
Wheeler set out to accomplish with his participatory universe as a
self-synthesizing system, providing a closed loop of existence (see
Figure 35).[66]

A self-explaining universe?

Taken on its own, then, a causal loop cannot account for *this particular* universe, because by invoking backward causation via time loops, quantum delayed choice, or any of the other physical mechanisms discussed by physicists, one can imagine a limitless number of possible self-creating universes. The vast majority will not resemble this one. Most of them will have no life, or perhaps life but no observers. So, clearly, invoking causal loops isn't enough by itself. The crucial additional feature in Wheeler's participatory principle is that the universe is not only self-creating, it also contains observers. But I don't think that observation alone is enough.

A clue to the missing ingredient comes from the fact that the laws of physics and the states of matter in this universe have the special property that they permit physical systems (brains, genes, computers) to construct an internal representation of the world – that is, to perform virtual reality computations that mirror the external universe. In short, they embody knowledge about the world. This self-reflective property of the universe is certainly part of the story, but only part. As I have stressed earlier in this chapter, for (some) human observers at least, the model of the world constructed by brains goes beyond mere knowledge of the world, it also includes interpretation and understanding. Through science and mathematics, we not only observe the drama of nature, but we have been able, albeit only partially so far, to unravel the plot, to glimpse the deep, hidden subtext of nature in the form of its subtle mathematical laws and principles, and to gain some understanding of how the universe is put together and works as a coherent system. So our universe possesses laws and states which not only permit self-simulation, they also permit *self-comprehension*. The cosmic rule book, in being fit for life and in facilitating the eventual emergence of consciousness, has not only ensured that the universe has constructed its own awareness. The cosmic scheme has also constructed an understanding of the cosmic scheme.

It may well be objected that the vast majority of people do *not* understand the cosmic scheme. In fact, few people have any real grasp of science or mathematics at all. And not even the most accomplished of scientists would claim to possess a *complete* understanding of the universe. But we must not be anthropocentric about this. Taking the optimistic view that intelligent life will survive and spread through the universe and gain an ever greater understanding of the workings of nature, it is easy to believe that, taken as a whole, the universe will in due course achieve complete self-understanding, though perhaps only through the final merging of mind and cosmos discussed in the previous section.

It seems obvious that a universe cannot create and explain itself without also understanding itself. If we take the metaphor of the universe as a computer, we can think of the product of scientific and intellectual endeavour as being the output of a 'cosmic program'. The input is made up of the laws as originally selected or generated (however that happened). It is well known that a computer algorithm cannot deliver more information in the output than is present in the input: the algorithm merely processes the information, it doesn't augment it. Consider the self-explanatory loop 'A explains B, B explains A'. If the word 'explanation' is to be interpreted in the sense of a sequence of logical steps (as in a computer algorithm), it then follows that A and B must have the *same* information content. I think an analogous argument can be used not merely for information content per se (just counting bits), but for something like the 'ingenuity' or 'meaningfulness' of the information. To invert the famous dictum 'garbage in, garbage out', I am claiming something like 'meaning out, meaning in'. If the universe runs on an ingenious cosmic code, and if the existence of the code is attributed to a self-consistent, self-explanatory loop, then the state of the universe has to be, at some point in its evolution, equally as ingenious as the laws that underpin it. The universe clearly cannot be self-explanatory without containing the ability to explain itself![67] If there is to be a complete explanation for the universe as a loop, the universe has to know *and understand* the laws it is responsible for in order to bring those laws into being. How could it be otherwise?

Existence explained? Some outstanding questions

In this final chapter I have outlined the basis of a scientific theory of existence which seeks to avoid appealing to a 'super-turtle', such as immutable transcendent laws, which must be accepted on faith alone. Whether or not such a project can be successfully completed is a matter for future researchers; there are many unanswered questions which need to be addressed. It is easy enough to state that the universe has engineered its own awareness through some sort of self-synthesizing principle, but how exactly is that achieved? And by what physical mechanisms? I have pointed out the importance of replacing the notion of rigid Platonic laws with quasi-laws that emerge, or crystallize, from the ferment of the big bang, but what exactly is the process that crystallizes them, and how is the selection made – or is it arbitrary, like spontaneous symmetry-breaking? How does one avoid a 'law of law selection', which would bring us right back to the problem of super-turtles? The answers must lie in the constraint of self-consistency. In his essay 'Computation and physics: Wheeler's meaning circuit?', Rolf Landauer made this point explicitly: 'Computation is a physical process ... Physical law, in turn, consists of algorithms for information processing. Therefore, the ultimate form of physical laws must be consistent with the restrictions on the physical executability of algorithms, which is in turn dependent on physical law.'[68] In other words, the laws determine what can be computed, and computability determines the laws. The open question is whether this requirement of self-consistency is enough to pin down the actual form of the laws. Is there only one self-consistent loop, or many? Are there an infinite number?

Landauer and Wheeler drew heavily on quantum mechanics in their analysis. Although quantum mechanics is a breathtakingly successful theory in its application, its interpretation remains confused and hotly debated. In seeking to construct his participatory universe scenario, Wheeler clung to the original, so-called Copenhagen interpretation of quantum mechanics, possibly because he collaborated as a young man with its Danish architect, Niels Bohr. In the Copenhagen interpretation (see Box 8, p. 257) the act of observation plays a central role.

On that basis, Wheeler claimed that only a universe containing observer-participators could exist – a version of the strong anthropic principle.

Most cosmologists, however, reject the Copenhagen interpretation in favour of the quantum multiverse, which describes an infinite ensemble of really-existing parallel universes. Some of these universes, or 'branches' of the quantum state, would contain life, and others wouldn't. The rules of quantum mechanics don't permit you to chop out the dead branches of the quantum state and discard them in some cosmic dustbin. So the price you pay for universes with life is the existence of a vast number of dead universes, a sort of supporting cast of cosmic characters which go unobserved but contribute crucially to the universe's life-encouraging capabilities. If one thus enlarges one's view of the universe to embrace the quantum multiverse, there is a lot of real estate out there which is indirectly necessary for our existence. In this respect it is similar to the observation that the universe we do see contains enormous chasms of dead, empty space, but were it not so the universe would be either too young or too hot or both to permit life on planets like Earth.

A further issue is how a self-synthesizing universe relates to the cosmological multiverse described in the earlier chapters of this book, based on ideas from string theory and inflation. In the theory of eternal inflation, the big bang is no longer the ultimate origin of all physical things. Rather, it is the beginning of the history of our 'bubble', or pocket universe, embedded in an eternally inflating superspace. So it may be more appropriate to develop a theory of a self-synthesizing multiverse rather than a self-synthesizing single universe. Here the recent work of Stephen Hawking and Thomas Herzog may be relevant.[69] They also consider, in the context of quantum cosmology, the backwards-in-time aspect of quantum observations: that what we choose to observe today helps to shape the nature of the universe in the remote past. Hawking and Herzog are reluctant to accept as 'given' the existence of a cosmological multiverse that evolves forwards in time from some well-defined past quantum state, as for example in the model of eternal inflation. Instead, they prefer to start from the present and construct all the alternative past quantum trajectories – the many different histories permitted within quantum uncertainty – that lead to it. Naturally, observers will select only histories consistent with life and observers, even if such histories are rare among the set of all possibilities.

Another set of questions concerns the final anthropic principle and the distant future of the universe. In this chapter I have painted a picture of life and mind expanding to saturate the universe over an immense duration of time. But is this consistent with the laws of physics? It turns out that only in a restricted class of cosmological models is it possible for life and mind to engulf the whole universe. A related issue is whether the universe is capable of processing an infinite amount of information. The answer boils down to how the expansion rate of the universe will change in the future. In Chapter 6, I explained how the discovery of dark energy implies that a cosmological event horizon may form and that the end state of the universe will be one of dark emptiness. If that is an accurate prediction, then, as a more careful analysis shows, the observable universe can only *ever* contain a finite amount of information. In effect, it will be a finite state machine. A system that can exist in only a finite number of physical states can support only a finite number of mental states: the god-like super-brain would enjoy only a limited number of experiences, thoughts, insights, and so on, and would be condemned to repeat them again and again, over and over. Many people find this a depressing prospect (and it is, of course, quite unlike the traditional notion of an infinite deity). By contrast, an unbounded information processing system would be able to experience perpetual novelty. From the scientific standpoint, the case is by no means open and shut, because it is far from clear that dark energy must remain strictly constant with time. If it should fall towards zero, however slowly, then the horizon size will grow, and the information processing capability of the universe will grow with it. While at any given moment the universe would still contain a finite amount of information, there would be no limit to how large it could become eventually.

So, how come existence? At the end of the day, all the approaches I have discussed are likely to prove unsatisfactory. In fact, in reviewing them they all seem to me to be either ridiculous or hopelessly inadequate: a unique universe which just happens to permit life by a fluke; a stupendous number of alternative parallel universes which exist for no reason; a pre-existing God who is somehow self-explanatory; or a self-creating, self-explaining, self-understanding universe-with-observers, entailing backward causation and teleology. Perhaps we have reached a fundamental impasse dictated by the limitations of the human intellect. I began this book by saying that religion was the first great systematic attempt to explain all of existence and that science is the next great attempt. Both religion and science draw their methodology from ancient modes of thought honed by many millennia of evolutionary and cultural pressures. Our minds are the products of genes and memes.[70] Now we are free of Darwinian evolution and able to create our own real and virtual worlds, and our information processing technology can take us to intellectual arenas that no human mind has ever before visited, those age-old questions of existence may evaporate away, exposed as nothing more than the befuddled musings of biological beings trapped in a mental straitjacket inherited from evolutionary happenstance. The whole paraphernalia of gods and laws, of space, time and matter, of purpose and design, rationality and absurdity, meaning and mystery, may yet be swept away and replaced by revelations as yet undreamt of.

Key points

- Conventional explanations run into a tower-of-turtles problem. Some scientists and philosophers have suggested self-consistent explanatory loops instead.
- Some scientists reject the orthodox, Platonic view of physical laws (idealized, infinitely precise mathematical relationships which transcend the physical universe).
- The laws of physics may not operate with infinite precision, because the universe has finite computational power.
- Teleology, or final causation, is taboo in orthodox science. The concept of a universe destined to bring forth life and observers is clearly teleological.
- Backward causation could provide a quasi-respectable scientific route to teleology.
- Wheeler's delayed-choice experiment describes a way in which

observers today help shape the nature of reality in the past, without being able to send information backwards in time.

- If the universe is self-explanatory, then it must evolve beings able to explain it.

Afterword: Ultimate explanations

COPLESTON: . . . But your general point then, Lord Russell, is that it's illegitimate even to ask the question of the cause of the world?

RUSSELL: Yes, that's my position.

COPLESTON: If it's a question that for you has no meaning, it's of course very difficult to discuss it, isn't it?

RUSSELL: Yes, it is very difficult. What do you say – shall we pass on to some other issue?

Debate between Fr F. C. Copleston and Bertrand Russell[1]

It is inevitable that any discussion that sets out to grapple with the ultimate questions of existence will eventually slide well beyond the comfort zone of most scientists and enter a realm of speculation which may seem outlandish. I thought it would be helpful to finish by summarizing the pros and cons of the various main positions I have been examining in the book. Each has distinguished scientists and philosophers prepared to argue for it.

A. The absurd universe

This is probably the majority position among scientists. According to this point of view, the universe is as it is, mysteriously, and it just happens to permit life. It could have been otherwise, but what we see is what we get. Had it been different, we would not be here to argue about it. The universe may or may not have a deep underlying unity, but there is no design, purpose or point to it all – at least none that would make sense to us. There is no God, no designer, no teleological principle, no destiny. Life in general, and human beings in particular, are an irrelevant embellishment in a vast and meaningless cosmos, the existence of which is an unfathomable mystery.

The advantage of this position is that it is easy to hold – easy to the point of being a cop-out. If there is no deeper meaning or scheme, there is no point in searching for one. In particular, there is no point in seeking links between life, mind and cosmos: according to this view, there *is* no connection, apart from the trivial one that life has emerged from the cosmos and mind has emerged from life, purely by accident. The disadvantage of the absurd universe view is that science cannot be expected to uncover new and deeper layers of order or further connections between natural phenomena. If there is no coherent scheme of things, then the success of the scientific enterprise to date is rendered totally enigmatic, and science can be pursued only with a completely unjustified faith that the methods used hitherto will continue to uncover reasonlessly existing order beneath the surface appearance of things. The fact that life exists, seemingly against vast odds, is attributed to an extraordinary accident. And appealing to luck, like appealing to miracles, is not a very satisfactory explanation. That life has evolved mind has to be accepted as another stupendous accident of history. The fact that some minds are capable of under-standing the universe is likewise either dismissed as yet another fluke, or tied to vague notions that brains have evolved to recognize patterns, and that – again for no reason – the deep patterns of physics and cosmology resemble the patterns of the everyday world on our planet (which in fact they mostly don't).

B. The unique universe

This point of view holds that there *is* a deep underlying unity in physics, and there is a mathematical theory 'out there' that will pull it all together if only we are smart enough to formulate it. It could be string/M theory or something else. Whatever it is, it will turn out to be founded on a deep mathematical principle which leaves no room for adjustment. All the laws of physics, all the parameters in the Standard Model, the various constants of nature, the existence of space and time with three and one dimensions respectively, the origin of the universe, quantum mechanics, relativistic spacetime and its causal properties – the whole shebang – will follow inexorably and inevitably from this final unified theory. It will truly be a theory of everything.

In the extreme version of this position, call it *B1*, the universe must exist necessarily as it is; it could not have been otherwise. There is a unique, self-consistent description of physical reality. If there is a God, then this being will have nothing to do, apart from perhaps 'breathing fire into the equations', because there are no choices to be made, no free parameters, no room for design. In the less extreme version of this position, *B2*, the universe could have been otherwise: there could be many unified theories describing different self-consistent realities, but the one being sought is simply *the* one that works, for no reason that can be discerned. To that extent, the existence of this particular universe is either a mystery, or it is absurd because there is no reason why this rather than that self-consistent reality is 'the one'. This point of view (*B2*) seems to be held by most physicists working on the unification programme and other aspects of fundamental physics, such as high-energy physics.

The advantage of the unique universe position is that it holds out the dream of a complete understanding of physical existence. Nothing is left unexplained; nothing of a fundamental nature is arbitrary or the result of chance, or needs fixing by an unknown designer. If there can be only one universe (*B1*), then the final theory would represent the greatest triumph of the human intellect. We would finally know the reason for existence: it *had* to be like this (or not exist at all). The disadvantage of *B2* is that, although a parameter-free unified theory that does the job would be in our possession, the ultimate question of 'why that theory?' could remain unexplained. Most scientists would, I think, settle for that – for not knowing the answer to the ultimate question of existence. They would proclaim 'It's a mystery!' and move on to something else. A disadvantage of both *B1* and *B2* is that the bio-friendliness of the universe is shrugged aside as an insignificant coincidence. Because the theory fixes everything, it is unexpected good fortune that this fix turns out to be consistent with life and mind (not to mention understanding).

C. The multiverse

A minority of scientists, but a growing one, now support the multiverse theory in one version or another. Modern cosmological models point strongly to the existence of a multiplicity of cosmic domains (e.g. bubble universes, pocket universes, variegated cosmic regions) as a natural and generic feature in which the big bang that gave birth to our universe is but one of many (probably an infinite number of) bangs generating a multiplicity of 'universes'. In addition, many theories that seek to unify physics predict some sort of variability in at least some of the constants of nature – parameters that enter into the Standard Model of particle physics – and in some of these theories there is variation in the form of the laws of low-energy physics too, opening the way for them to vary from one cosmic domain to another as the universes cool from their melting-pot origins. The favoured unification model, or models, known as string/M theory, seems to entail a 'landscape' of vastly many possible low-energy universes, with nothing obvious to single out a special one.

The advantage of the multiverse theory is that it provides a natural and easy explanation of why the universe is so uncannily fine-tuned for life: observers arise only in those universes where, like Goldilocks' porridge, things are by accident 'just right'. Bio-hostile universes overwhelmingly proliferate, but they are by definition sterile, so they go unseen. The disadvantage of the multiverse theory is that it invokes an overabundance of entities, most of which could never be observed, even in principle. This profligacy strikes many people as an extravagant way to explain bio-friendliness. The theory is also very hard to test. Observers are treated simply as selection agents, so the mysterious comprehensibility of the universe (to the human mind at least) is left unexplained. The multiverse does not provide a complete account of existence, because it still requires a lot of unexplained and very 'convenient' physics to make it work. For example, there has to be a universe-generating mechanism, quantum mechanics has to describe everything, and unified laws of some sort (such as arise from string/M theory) have to be simply accepted as 'given'. So the multiverse, at least in this 'mild' form, lacks the power of B1 (the unique universe), although it is no worse than B2. Some sort of ingenious selection still has to be made, not of a universe but of a multiverse. The problem of existence has therefore not gone away, only been shifted up one level.

The last criticism is avoided by the extreme multiverse model proposed by Max Tegmark in which all possible worlds of any description really exist, not just those flowing from a specific mathematical model such as string/M theory and inflation. The advantage of the extreme multiverse is that it explains everything because it contains everything. This has the virtue of simplicity and 'naturalness', but the huge disadvantage of appearing rather vacuous. A theory which can explain anything at all really explains nothing. However, a multiverse which contains less than everything implies a rule that separates what exists from what is possible but does not exist. The rule remains unexplained. Another disadvantage of all multiverse theories is that they seem to lead to the prediction of fake universes which (at least on a simple counting basis) outnumber the real ones, leading to the bizarre conclusion that the observed universe is probably a fake, and so its physics cannot be taken seriously anyway.

Multiverse proponents get sniped at from both sides. Religious adherents regard the theory as a frantic attempt to dodge any sort of god: 'the last resort for the desperate atheist', in the words of the philosopher Neil Manson.[2] String/M theory purists, on the other hand, see it as a weak-kneed abdication of professional responsibility in the face of mathematical difficulties.

D. Intelligent design

The traditional monotheistic religious view is that the universe is created by God and designed to be suitable for life because the emergence of sentient beings is part of God's plan. This has the advantage of being a simple explanation of the cosmic fine tuning and bio-friendliness, and of being a 'natural' explanation for those people who have already decided on other grounds that God exists. It also attributes the design-like qualities of the universe to a designer, which seems reasonable enough. However, it suffers from the obvious disadvantage of being a conversation-stopper. The simple declaration 'God did it!' provides no actual explanation for anything, unless one can also say *how* and *why* God did it. It also runs into the problem of who designed the designer, unless the notion of a necessary being can be firmly established, and shown to be different from, and superior to, a necessary universe (in the sense of *B1*).

The other main problem with intelligent design is that the identity of the designer need bear no relation at all to the God of traditional monotheism. The 'designing agency' can be a committee of gods, for example. The designer can also be a natural being or beings, such as an evolved super-mind or super-civilization existing in a previous universe, or in another region of our universe, which made our universe using super-technology. The designer can also be some sort of superdupercomputer simulating this universe. So invoking a super-intellect as the levitating super-turtle is fraught with problems.

E. The life principle

In this theory, the bio-friendliness of the universe arises from an overarching law or principle that constrains the universe/multiverse to evolve towards life and mind. It has the advantage of 'taking life seriously', treating it neither as a completely unexplained bonus, as in *A* and *B*, nor as a mere passive selector, as in *C*. It avoids the 'gerrymandering' feel of *D*, replacing a manipulative (natural or supernatural) god with a more subtle, purpose-like principle. In short, it builds purpose into the workings of the cosmos at a fundamental (rather than an incidental) level, without positing an unexplained pre-existing purposive agent to inject purpose miraculously.

The disadvantage is that teleology represents a decisive break with traditional scientific thinking, in which goal-oriented or directional evolution is eschewed as anti-scientific. Critics ask how the universe 'knows' about life in order to contrive its eventual emergence. This raises the problem of causation, both of how to accommodate an additional life principle in a system of physical laws which is already supposed to do the job of explaining everything, and also the weirdness of backwards-in-time causation, or backwards-in-time something. As I have explained, these may not be fatal flaws, but they certainly make scientists nervous. Atheistic scientists regard any talk of directional principles as a cover for the guiding hand of God being slipped back into science, even if it is a far cry from the God of traditional monotheism. A life principle also suffers from the problem of singling out life and mind as the 'aim' of cosmic evolution, without explaining why. One could just as well nominate any distinctive and complex state of matter and enshrine its emergence in a teleological principle. So the life principle itself must just be accepted as a brute fact, along with the laws of physics, existing without any explanation. This objection is readily removed if one combines a teleological principle with the multiverse, because only universes with *life* principles built into their laws get a chance to be observed. But invoking the multiverse merely transfers the problem of where the life principle came from to the problem of where the multiverse came from.

F. The self-explaining universe

All the foregoing options hit the tower-of-turtles problem, with the exception of *B1*, the Tegmark version of the multiverse (under *C*), and the existence of a necessary God (under *D*). Something un-explained has to be accepted as given and the rest of the explanatory scheme constructed on that ad hoc foundation. One way to avoid this trap is to appeal to a closed explanatory or causal loop. In effect, the universe (or multiverse – it can work at both levels) explains itself. There are even models involving causal loops or backwards-in-time causation, where the universe creates itself. The advantage of such a scheme is that it is self-contained and avoids both the infinite regress of the tower of turtles, and the act of faith involved in invoking a levitating super-turtle. The disadvantage is that we are still left not knowing why this universe – *this* self-explaining, self-creating system – is the one that exists, as opposed to all other self-explanatory schemes. Perhaps *all* self-explanatory schemes exist and only ones like ours get observed because they are consistent with life – another variant on the multiverse. Or, better still, perhaps existence isn't something that gets bestowed from outside, by having 'fire breathed' into a potentiality by some unexplained fire-breathing agency (i.e. a transcendent existence generator), but is also something self-activating. I have suggested that only self-consistent loops capable of understanding themselves can create themselves, so that only universes with (at least the potential for) life and mind really exist.

G. The fake universe

We are living in a simulation, and what we take to be the real world is an ingeniously contrived virtual reality show. This is a variant on the intelligent designer scenario, but upgraded for the information age. This theory enjoys the same easy-fix advantages as intelligent design, but has the distinct disadvantage of undermining the scientific quest. If the universe is a sham, why bother to figure out how it works?

H. None of the above

Did I leave anything out?

My own inclinations, it will be clear, lie in the directions of E and F, although there are many details to be worked out. I do take life, mind and purpose seriously, and I concede that the universe at least *appears* to be designed with a high level of ingenuity. I cannot accept these features as a package of marvels which just happen to be, which exist reasonlessly. It seems to me that there is a genuine scheme of things – the universe is 'about' something. But I am equally uneasy about dumping the whole set of problems in the lap of an arbitrary god, or abandoning all further thought and declaring existence ultimately to be a mystery.

It is often argued that science is, or should be, value free. Certainly science, conducted properly, is the realm of human inquiry least tainted by preconceived prejudice and ideology. But scientists (me included) will inevitably formulate opinions which draw on a more general world view, incorporating personal, cultural and even religious elements. Many scientists will criticize my E/F inclination as being crypto-religious. The fact that I take the human mind and our extraordinary ability to understand the world through science and mathematics as a fact of fundamental significance betrays, they will claim, a nostalgia for a theistic world view in which humankind occupies a special place. And this even though I do *not* believe *Homo sapiens* to be more than an accidental by-product of haphazard natural processes. Yet I do believe that life and mind are etched deeply into the fabric of the cosmos, perhaps through a shadowy, half-glimpsed life principle, and if I am to be honest I have to concede that this starting point is something I feel more in my heart than in my head. So maybe that is a religious conviction of sorts.

People of a more mainstream religious persuasion will regard D as self-evidently true, and dismiss my attempt to go beyond the traditional God as a sign that I have succumbed to the indoctrination of scientism. Those scientists who, by contrast, passionately hope for B usually make no bones about the fact that they are committed to a form of ideology. I do not begrudge the unifiers their chance; if they can pull off true unification, it will be not just the greatest scientific theory of all time, but the theory to end all theories. Yet the hostility of some of them to C (the multiverse) and D, E, F does carry the hallmarks of an extra-scientific agenda. There is also a sizeable group of scientists who, perhaps in reaction to the homocentrism of traditional religion, or motivated by dismay at humanity's brutality and destruction of the environment, wish to diminish or even besmirch human significance, and with it the significance of human qualities such as intelligence and understanding. For these scientists, any suggestion of a teleological trend or progressive evolution towards consciousness, or even towards greater complexity, is anathema. Their arguments, however, also carry barely concealed overtones of an ideological agenda. In this respect they are little different from those who have decided in advance on this or that religious interpretation of nature, and then shoehorn the scientific facts to fit their preconceived beliefs. Meanwhile, it has to be admitted, most scientists stick with something like position A and get on with their work, leaving the big questions to philosophers and priests.

Notes

Preface and Acknowledgements

1 B. J. Carr and M. J. Rees, 'The anthropic principle and the structure of the physical world', *Nature*, vol. 278 (1979), p. 605.

2 John Barrow and Frank Tipler, *The Anthropic Cosmological Principle* (Oxford University Press, Oxford, 1986).

3 Martin Gardner, 'WAP, SAP, PAP, & FAP', *New York Review of Books*, 8 May 1986.

4 Leonard Susskind, *The Cosmic Landscape: String Theory and the Illusion of Intelligent Design* (Little Brown, New York, 2005), p. 138.

5 Bernard Carr (ed.), *Universe or Multiverse?* (Cambridge University Press, Cambridge, 2006).

6 Paul Davies, *The Mind of God* (Simon & Schuster, London, 1992).

1: The Big Questions

1 I shall restrict my discussion to life as we know it. The possibility of exotic forms of life based on other chemical elements, or other physical processes entirely, is certainly fascinating but completely speculative. If life is common, we have no reason to suppose that our form of life is atypical. Readers interested in a less conservative approach will find an up-to-date discussion in Peter Ward, *Life As We Do Not Know It* (Viking, New York, 2005).

2 Fred Hoyle, 'The universe: past and present reflections', *Annual Review of Astronomy and Astrophysics*, vol. 20 (1982), p. 16.

3 See, for example, David Park, *The Grand Contraption: The World as Myth, Number and Chance* (Princeton University Press, Princeton, NJ, 2005).

4 The term was popularized by the physicist Heinz Pagels in *The Cosmic Code* (Simon & Schuster, New York, 1982).

5 See, for example, Edward Craig, *The Mind of God and the Works of Man* (Oxford University Press, Oxford, 1987).

6 See, for example, John W. Carroll, *Laws of Nature* (Cambridge University Press, Cambridge, 1994); and Alan Padgett, 'The roots of the Western concept of "Laws of Nature": from the Greeks to Newton', *Perspectives on Science and Christian Faith*, vol. 55, no. 3 (December 2003), p. 212.

7 Lucretius, *De rerum natura*, edited by M. F. Smith (Hackett Publishing Co., Indianapolis, 2001), p. 138.

8 Marcus Manilius, *Astronomica*, translated by G. P. Goold (Heinemann, London, 1977), p. 121.

9 Augustine, *The Literal Meaning of Genesis*, Vol. 2, translated by J. H. Taylor (Paulist Press, New York, 1983), p. 92.

10 Stillman Drake, *Discoveries and Opinions of Galileo* (Doubleday-Anchor, New York, 1957), p. 70.

11 James Jeans, *The Mysterious Universe* (Cambridge University Press, Cambridge, 1930), p. 140.

12 Unbeknownst to me, Lindsay's naive question was being asked at about the same time by one of the world's leading theoretical physicists, Eugene Wigner: see his 'The unreasonable effectiveness of mathematics in the natural sciences', *Communications in Pure and Applied Mathematics*, vol. 13, no. 1 (1960), p. 1.

13 There is a minority school of thought which says that this is all baloney, that the laws of physics are just human inventions constructed for convenience, and that there are no 'real' laws at all. I am going to ignore this dissenting position because I think it is totally wrong and does not merit serious discussion.

14 Nancy Cartwright, *How the Laws of Physics Lie* (Clarendon Press, Oxford, 1983).

15 David Mowaljarli and Jutta Malnic, *Yorro Yorro* (Magabala Books, Broome, Australia, 1993), chapter 23.

16 Richard Feynman, 'The meaning of it all', 1963 John Danz Lecture, published under the same title by Addison Wesley (Reading, MA, 1998), p. 14.

17 Steven Weinberg, *The First Three Minutes* (André Deutsch, London, 1977), p. 149.

2: The Universe Explained

1 K is the symbol for the unit of temperature called the Kelvin. A temperature interval of one degree Kelvin is the same as an interval of one degree Celsius, but the Kelvin scale starts from absolute zero, or about –273°C.

2 Roughly speaking, this is the final state into which a closed system settles, following which no large-scale changes occur. For a simple gas, it is a state of uniform pressure and density.

3 An ionized gas, also called a *plasma*, is one in which the atoms are dissociated into electrons and nuclei, as is caused by extreme heat. I shall describe the primordial gas in more detail in Chapter 3.

4 The subsequent scattering of the CMB from early clumps of gas produced subtle effects in the polarization of the radiation, effects which have also been detected by WMAP.

5 Reproduced by kind permission of the correspondent.

6 Sometimes cosmologists refer to 'the last scattering surface', the spherical shell of matter surrounding Earth from which the radiation emanates at the moment of transition from opaque to transparent. This transition is technically termed the 'decoupling' of matter and radiation.

7 For a careful exposition of this point, see Tamara Davis and Charles Lineweaver, 'Misconceptions about the big bang', *Scientific American* (March 2005), p. 36.

8 This issue is complicated by the theory of inflation, which I shall describe in Chapter 3.

9 Analogously, when a ship disappears over the terrestrial horizon, we do not infer that the Earth ends there.

10 This admirable term was suggested by Alan Guth, and I have decided to adopt it here.

11 Time is not a dimension of space, but a dimension of spacetime.

12 This won't apply if theories about 'branes' turn out to be correct – see p. 54.

13 To sound a note of caution, some cosmologists are concerned that the largest features mapped by WMAP (technically, the lowest multipoles), display some oddities not predicted by the conventional big bang model of the universe. It is too soon to know whether this is due to problems with the equipment and/or data analysis, or points to something significant and unexpected about the structure of the universe.

14 The limited accuracy of these observations cannot establish that the universe is *exactly* flat. What they tell us is that if the universe is shaped like Einstein's hypersphere, then the radius of the hypersphere is

exceedingly large, so that within the volume of space probed by our instruments we cannot discern any curvature. Similar remarks apply to any negative curvature.

15 Even if space is flat, it need not necessarily be infinite. That is because Einstein's theory says nothing about the *topology* of space. One possible topology involves identifying points. Think of a sheet of paper on which a particle enters from the left, traverses the paper, and exits from the right. Now imagine rolling up the paper and gluing the left and right edges together. The particle that previously exited from the right would now reappear from the left. Some cosmologists have suggested that space might be like this, and resemble a hall of mirrors. If we inhabited such a universe it might look to us at first sight as if 'the hall of mirrors' extended to infinity, but on closer inspection we would discover that a finite volume of space repeats itself, infinitely often. It is possible that the universe consists of three-dimensional cells, repeated periodically, and that light which we take to be from far away is in fact wrapping around one or more times, creating the illusion of distance. More complicated shapes, such as the three-dimensional analogue of the surface of a segmented soccer ball, have also been suggested.

16 I am being a little cavalier with my terminology. The word 'matter' here includes both dark matter and dark energy, topics I shall discuss in Chapter 6.

17 *Flatland: A Romance of Many Dimensions* by E. A. Abbott is now available in an annotated edition by Ian Stewart (Perseus Publishing, New York, 2001).

18 See, for example, Lisa Randall, *Warped Passages* (Allen Lane, London, 2005).

19 G. J. Whitrow, 'Why space has three dimensions', *British Journal for the Philosophy of Science*, vol. 6, no. 21 (1955), p. 1.

3: How the Universe Began

1 The helium that is used to fill balloons comes not from the big bang, but from the product of radioactive decay in the Earth's interior.

2 This 'happy medium' is related to the fact that space is flat.

3 Inflation was originally invented by Guth to solve a different problem – the absence of entities known as magnetic monopoles. Alan Guth's own account can be read in his book *The Inflationary Universe: The Quest for a New Theory of Cosmic Origins* (Perseus Publishing, New York, 1998).

4 The word 'scalar' means that the field can be described simply by specifying

a single number (the strength of the field) at each point in space. By contrast, an electric field has both a magnitude *and* a direction at each point; it is a so-called *vector field*. Gravitation is more complicated still – a *tensor field* – requiring even more numbers at each point to fully describe it.

5 Don't confuse the mechanical force exerted by the pressure, which is huge and outward (though contained by Earth) with the gravitational force that this pressure generates, which is tiny and inward.

6 Pressure and energy are normally measured in different units. To discuss the correspondence between these quantities you must divide the pressure by c^2, which then gives it the same units as energy density. This large divisor explains why energy gravitates so much more strongly than pressure.

7 Mechanically, the scalar field sucks – fiercely; gravitationally it repels – gently. You might be wondering why, if this scalar field sucks so much, it doesn't pull itself into a smaller and smaller region. That is because it is spread uniformly through space, so there is no privileged place for it to converge: it is being sucked every way at once, and there is no net force to pull it to any particular place.

8 There is, however, a further issue about the creation of matter, related to the question of antimatter. I shall defer this complication until the next chapter.

9 Good popular accounts have been written by some of the originators. In addition to Guth's book, see, for example, Andrei Linde, *Inflation and Quantum Cosmology* (Academic Press, San Diego, Calif., 1990).

10 Particle creation by the expansion of the universe is a purely gravitational (and normally very weak) process. It should not be confused with particle production from the decay of the inflaton field, or from heat energy (such as occurred at the end of inflation).

11 'Ex nihilo, nihil fit'. *De rerum natura* in Lucretius, *On the Nature of the Universe*, translated by R. E. Latham (Penguin, London, 1951).

12 In Chapter 10, I shall consider the extremely speculative idea of back-wards-in-time causation, where the big bang could be said to have been caused by later events retroactively.

13 *City of God*, Book xi.6, in *Basic Writings of St. Augustine*, edited by W. T. Oates (Random House, New York, 1948).

14 It is sometimes conjectured that in the cyclic model the state of the universe is somehow reset at the bounce (technically, the entropy is reduced). However, this step is rather contrived. It either has to be imposed by hand, or tied to a more complicated – and speculative – model of the sort I explain in Chapter 10.

15 For the mathematically inclined, the Planck length is given by $(Gh/2\pi c^3)^{1/2}$.

16 You may wonder why quantum effects of electromagnetism set in at

atomic dimensions, whereas quantum effects of gravitation are predicted to be important only on much smaller scales of size. The reason stems in part from the huge disparity in strength between the two forces, a topic I shall discuss in the next chapter.

17 This is known technically as the 'no-boundary' proposal.

18 Hawking's own account can be found in his book *A Brief History of Time* (Bantam, New York, 1988).

19 John Leslie, *Universes* (Routledge, London, 1989), p. 95.

20 This is a curious inversion of the usual situation in quantum mechanics. In the inflating universe, the most conspicuous consequences of quantum mechanics are on the *largest* scale of size.

21 In this respect, eternal inflation is reminiscent of the old steady-state theory of cosmology, championed by Hoyle, in which the universe has no beginning or end, but new matter is continually created as the universe expands so as to maintain an unchanging average density. Where eternal inflation differs is that entire pocket universes are created rather than particles of matter.

22 A. Linde, 'Inflation and quantum cosmology', in *300 Years of Gravitation*, edited by S. W. Hawking & W. Israel (Cambridge University Press, Cambridge, 1987), p. 618.

23 Leonard Susskind, *The Cosmic Landscape: String Theory and the Illusion of Intelligent Design* (Little Brown, New York, 2005), chapter 11.

24 This is something of a simplification. When using the theory of relativity, we have to remember that distances, like times, are not absolute but relative, so we must always specify the circumstances of the observer when discussing a distance. Paradoxically, if the observer is located inside one of the bubbles (as we are within our pocket universe), it is possible for the size of the bubble to be *infinite* relative to that observer, even though, viewed from outside the bubble, it is finite.

25 David Hume, *Dialogues Concerning Natural Religion*, edited by Martin Bell (Penguin, London, 1990), part V, p. 77.

4: What the Universe Is Made of and How It All Holds Together

1 Even uranium plays a role in life on Earth. Its slow radioactive decay over billions of years keeps the interior of our planet hot, driving the convection currents that move the continental crust around, an essential process for recycling carbon and other substances used to maintain our ecosystem.

2 Positrons are today familiar from their role in medical imaging in the form of positron emission tomography (PET) scans.

3 These decay schemes also involve neutrinos.

4 When heavy particles decay into lighter ones, the excess mass-energy appears in the form of kinetic energy: the decay products are created moving at high speed.

5 Why stop there? Perhaps quarks (and maybe leptons too) are made out of yet smaller particles, which are in turn made of even smaller particles, and so on. Such ideas have been tried. But most physicists think that quarks and leptons are the bottom level, in terms of composite particle combinations. That may not be the last word, however, as I shall discuss at the end of this chapter.

6 The masses of the neutrinos are still being worked out. They all seem close to zero.

7 The stability of neutrinos is more complicated. They don't decay as such: instead, they keep rotating their identities between different neutrino flavours.

8 The word 'recoil' is a bit misleading here, because if the charges were of opposite sign, the deflection would be inward rather than outward. As a result of Heisenberg's uncertainty principle, the momentum transfer can be negative in quantum processes, causing an inward jerk rather than an outward deflection. However, the general picture in terms of virtual photon exchange is the same.

9 Mathematically speaking, one integrates over a weighted set of possibilities.

10 This procedure is known as perturbation theory.

11 This statement refers to the photon's *rest* mass (see Box 1, p. 36).

5: The Lure of Complete Unification

1 How might a process that takes on average much longer than the age of the universe show up in an experiment? The answer lies with the statistical nature of quantum mechanics. There is a certain probability that, from among a huge number of protons (many tonnes of material), one or two protons will decay in, say, a month. The experimenters looked for such occasional isolated decay events, but saw nothing.

2 It is important to understand that the particles emanating from high-energy collisions are not just the constituents of the impacting bodies: many of them are created *ab initio* from the energy of impact. For example, physicists routinely create electron–positron pairs, or proton–antiproton pairs.

3 The link between the spins of particles and the collective properties of assemblages of them as governed by the Pauli exclusion principle is not obvious, and has to do with certain abstract symmetries involved in the quantum concept of spin.

4 This law has the same general form as gravitation, as shown in Figure 1 (p. 10).

5 The technical term given to this difficulty is 'non-renormalizability'.

6 It does have something to say about the ultra-hot, very early universe though, and it is not impossible that some stringy relic may be found by cosmologists. But so far there is no sign of any.

7 The problem of multiplicity is greatly exacerbated by the existence in the theory of so-called fluxes, analogous to lines of electric or magnetic force, which can thread through the compactified spaces in a colossal number of different ways.

8 Amazingly, the idea of 'an extensible model of an electron' as a membrane was introduced into theoretical physics as long ago as the 1960s, by Paul Dirac. In the 1980s the class of extended objects was generalized from strings and membranes to any number of higher dimensions that is less than the dimensionality of the space in which they moved. This wider class became known as p-branes. The early history of branes is reviewed by Michael Duff in 'Benchmarks on the brane' (hep-th/0407175; 23 February 2005).

9 Polchinski called these membranes D-branes, as distinct from p-branes, and like p-branes they can be generalized to 3, 4, etc. dimensions.

10 Michio Kaku, 'Unifying the universe', *New Scientist*, 16 April 2005.

6: Dark Forces of the Cosmos

1 'Light elements' is the term used to mean the lowest-mass elements. They include deuterium – which, confusingly, is also known as 'heavy hydrogen'.

2 The word 'massive' here means 'high mass': it does not mean large in physical size. WIMPs would be point-like particles, but individually weighing more than the heaviest atoms.

3 An excellent account of dark matter in its different forms is given by Joel Primack and Nancy Abrams, *The View from the Centre of the Universe* (Fourth Estate, London, 2006).

4 Stephen Baxter, *Time* (HarperCollins, London, 1999).

5 P. C. W. Davies, 'Cosmological event horizons, entropy and quantum particles', *Annales de l'Institut Henri Poincaré*, vol. 49, no. 3, (1988), p. 297.

6 Robert R. Caldwell, Marc Kamionkowski and Nevin N. Weinberg, 'Phantom energy: dark energy with $w < -1$ causes a cosmic doomsday', *Physical Review Letters*, vol. 91 (2003), 071301–1.

7 Freeman Dyson, 'Time without end: physics and biology in an open universe', *Reviews of Modern Physics*, vol. 51, no. 3 (1979), p. 447.

8 It is possible that a supercivilization could engineer a new 'baby' universe as an escape route: see Chapter 8.

7: A Universe Fit for Life

1 Nicolaus Copernicus, *De revolutionibus orbium coelestium* ('On the Revolutions of the Heavenly Spheres') (modern printing by Prometheus Books, Amherst, 1995), p. 8. Originally published in Nurenburg, 1543.

2 The anthropic principle has a large literature. A comprehensive treatment with many references is given by John Barrow and Frank Tipler, *The Anthropic Cosmological Principle* (Oxford University Press, Oxford, 1986).

3 B. Carter, 'Large number coincidences and the anthropic principle in cosmology', in *Confrontation of Cosmological Theories with Observational Data*, IAU Symposium 63, edited by M. Longair (Reidel, Dordrecht, 1974), p. 291.

4 See, for example, Paul Davies, *The Origin of Life* (Penguin, London, 2003). Actually, it is rather more favourable for the transfer to occur the other way – that is, for life to start on Mars and come to Earth inside ejected rocks. Either way, one would still be dealing with a single genesis event.

5 Some science fiction writers, and a few scientists, have speculated about life based on very different chemical or physical processes, and it's true that scientists have no clear idea of what might or might not be possible. Even harder to assess are the possibilities for life based on radically different laws of physics. I shall adopt the conservative position that, in the absence of evidence to the contrary, life is restricted to something close to what we know.

6 I shall discuss only a handful of examples. Readers wanting a more complete treatment should refer to Barrow and Tipler, *The Anthropic Cosmological Principle*.

7 As a result, neutrinos are emitted. The neutrino flux from the sun has been measured with very sensitive equipment. Neutrinos have extremely low (rest) mass. If that were not the case, protons would lack the necessary mass-energy to turn into neutrons inside stars, thus preventing the sun from shining steadily and sustaining life.

8 For full details, see Richard Dawkins, *The Ancestor's Tale* (Weidenfeld & Nicolson, London, 2005).

9 Lithium and beryllium get manufactured as by-products of other reactions.

10 H. Oberhummer, A. Csótó and H. Schlattl, 'Stellar production rates of carbon and its abundance in the universe', *Science*, vol. 289 (2000), p. 88.

11 The word 'ylem' is an obsolete Middle English word meaning the primordial substance from which matter formed. Gamow used the term to mean a mixture of protons and neutrons.

12 Tritium is an isotope of hydrogen with nuclei containing *two* neutrons and one proton, so it is even heavier than deuterium.

13 By this, Gamow means a nucleus with either two protons and three neutrons or three protons and two neutrons. As I have mentioned, neither configuration is stable.

14 G. Gamow, *My World Line: An Informal Biography* (Viking, New York, 1970), p. 127. Reprinted courtesy of the Gamow Family Estate.

15 More familiar is the decay of a neutron into a proton, with an attendant release of an antineutrino. The reverse process I am discussing here, with a proton turning into a neutron, can happen in an imploding star because the intense gravitational field that is created supplies the necessary energy.

16 There are other, less efficient, ways for stars to divest themselves of carbon, so it is not clear how critical the neutrino interaction strength is to the fine-tuning argument for this element.

17 Neutron decay is a statistical process subject to quantum fluctuations. Half-life is defined as the average time it takes for exactly half of a population of neutrons to decay.

18 Max Tegmark, Anthony Aguirre, Martin Rees and Frank Wilcek, 'Dimensionless constants, cosmology and other dark matters', *Physical Review D*, vol. 77 (2006), p. 23505.

19 That is, why is the model so wrong – I don't think we made a mistake in our sums!

20 Inflation requires dark energy to be non-zero for a very brief time just after the big bang, but physicists still assumed that in the post-inflation phase the dark energy would drop to precisely zero.

21 Leonard Susskind, *The Cosmic Landscape* (Little Brown, New York, 2005), p. 78.

22 As far as I know, Sydney Coleman of Harvard University, who helped to pioneer the subject of symmetry-breaking in the early universe, was the first person to use the term 'the big fix' to describe the dramatic suppression of dark energy.

23 Steven Weinberg, 'Anthropic bound on the cosmological constant', *Physical Review Letters*, vol. 59 (1987), p. 2607.

24 The formation of galaxies depends delicately on the magnitudes of both the dark energy and the primordial density fluctuations. In my discussion I am assuming that the latter is held fixed while the former is allowed to vary. If both quantities are allowed to vary together, the analysis is more complicated. See, for example, Tegmark et al., 'Dimensionless constants, cosmology and other dark matters'.

8: Does a Multiverse Solve the Goldilocks Enigma?

1 That solitary individual was I. L. Rozenthal, who succeeded in publishing a credible review paper (*Soviet Physics Uspekhi*, vol. 23 (1980), p. 296). This was no mean feat in a regime which strongly discouraged any discussion that departed from the strict Marxist philosophy of dialectical materialism.

2 The various constants I have mentioned assume numerical values which depend on the system of units used to express them. For example, the speed of light is either (roughly) 300,000 km per second or 186,000 miles per second. Constants may be combined to form dimensionless ratios, which are pure numbers, independent of units. For example, the square of the charge on the electron divided by Planck's constant and the speed of light is a pure number with a value close to 0.001617. When considering whether the laws of physics contain free parameters that might vary from place to place, it makes sense only to discuss variations of such dimensionless ratios.

3 Neutrinos fall outside this scheme. Experiments show that they do have a tiny mass, but its explanation lies beyond the Standard Model.

4 The Higgs particle is a boson because it has spin 0.

5 James Watson, *The Double Helix* (Touchstone, New York, 2001).

6 This example can be likened to the rule of the road. In some countries people drive on the right, in others they drive on the left. Which one is chosen is just a matter of historical accident. It doesn't make any difference so long as everybody uses the same rule.

7 If you did the experiment *very* precisely, the selection of the direction could be traced back to chaotic molecular jiggles.

8 This example is due to Sidney Coleman.

9 By 'low-temperature' and 'low-energy' I mean low compared with the temperature and energy of symmetry-breaking. As we shall see, that may involve GUT or even Planck values. Given these elevated scales, what physicists normally refer to as 'high-energy physics' is still very low-energy indeed. So the low-energy world includes the world of subatomic accelerators such as the LHC, as well as everyday experience.

10 The alert reader may notice that this is about the time when inflation is supposed to have happened – which is no coincidence. It was by considering the application of GUTs to the very early universe that Alan Guth got the idea of the inflationary universe scenario in the first place, and in fact a plausible candidate for the inflaton field is one of the GUT Higgs fields.

11 Actually, I'm making this up. Nobody knows, because the theory is too complicated. But there are lots of options.

12 Leonard Susskind, *The Cosmic Landscape: String Theory and the Illusion of Intelligent Design* (Little Brown, New York, 2005), p. 21.

13 The existence of a landscape is based on a consideration of the five 'corners' of M theory representing the five original string theories, which can be studied using an approximation method called perturbation theory. Some theorists believe that the landscape is an artefact of this approximation, and predict that if the full underlying M theory could be properly formulated and solved exactly, it would yield a single, unique description – just one world. I shall have more to say about the alternative view in Chapter 9.

14 The idea that eternal inflation might provide a natural mechanism to generate large cosmic domains (pocket universes) with very different low-energy physics, and with obvious anthropic consequences, dates from the early 1980s. See A. D. Linde, 'The new inflationary universe scenario', in *The Very Early Universe*, edited by G. W. Gibbons, S. W. Hawking and S. Siklos (Cambridge University Press, Cambridge, 1983), p. 205. For an up-to-date account of this 'landscape exploration' process, see chapter 11 of Susskind's book *The Cosmic Landscape*.

15 The theories I have described here are by no means the only idea for a multiverse. A list of various multiverse theories has been compiled by Nick Bostrom in *Anthropic Bias: Observations and Selection Effects* (Routledge, New York, 2002); see also John Leslie, *Universes* (Routledge, London, 1989).

16 An excellent in-depth discussion and critique of these issues can be found in Neil Manson (ed.), *God and Design* (Routledge, London, 2003).

17 *The Edge* annual question, 2006. See http://www.edge.org

18 This type of reasoning is fully convincing only if one can assign precise statistical weights to different universes, but we don't know how to do that yet. Another assumption is that there is no obvious minimum value of the dark energy below which life would be impossible, unless one considers negative values. A substantial amount of negative dark energy would be life-threatening for a different reason: it would add to the gravitational attraction of the universe, and cause rapid collapse to a big crunch.

19 More details of this work can be found in John Barrow, *The Constants of Nature* (Jonathan Cape, London, 2002).

20 Max Tegmark, 'Parallel universes', *Scientific American* (May 2003), p. 31.

21 There is also a hidden assumption that the systems being considered have a finite, albeit very large, number of possible states. This is the case for discrete variables, as arise from the application of quantum mechanics, but there is no logical reason why some physical variables should not be continuous. If that were so, there would be infinitely many 'shades of grey', and the question of truly identical copies would be more subtle.

22 Nick Bostrom, 'The simulation argument: why the probability that you are living in a matrix is quite high', *Times Higher Education Supplement*, 16 May 2003. For a more scholarly analysis see his 'Are you living in a computer simulation?', *Philosophical Quarterly*, vol. 53, no. 211 (2003), p. 243.

23 The assumption that all physical processes can in principle be simulated by a universal computer rests on an unproven but widely believed conjecture called the Church–Turing thesis (named after Alan Turing and the American logician Alonzo Church). See, for example, David Deutsch, *The Fabric of Reality* (Allen Lane, London, 1997), p. 134.

24 Cited in J. R. Newman, *The World of Mathematics* (Simon & Schuster, New York, 1956).

25 A collection of essays on this topic can be found in Daniel Dennett and Douglas Hofstadter, *The Mind's I* (Harvester, Brighton, 1981). See also David Chalmers, '*The Matrix* as metaphysics', in *Philosophers Explore The Matrix*, edited by Christopher Grau (Oxford University Press, Oxford, 2005).

26 Alan Turing, 'Computing machinery and intelligence', *Mind*, vol. 59 (1950), p. 433.

27 A classic being Isaac Asimov's *I, Robot* (Genome Press, New York, 1950).

28 Roger Penrose, *The Emperor's New Mind* (Oxford University Press, Oxford, 1989).

29 Gordon Moore, co-founder of Intel, predicted decades ago that computing power would double every one or two years. So far he has been proved correct.

30 See, for example, Frank Tipler, *The Physics of Immortality* (Doubleday, New York, 1994).

31 Interested readers can learn more by visiting Bostrom's website at www.simulation-argument.com

32 Martin Rees, *Our Final Century* (Random House, London, 2003).

33 John Barrow, 'Glitch', *New Scientist* (7 June 2003), p. 44. Reprinted courtesy of *New Scientist*.

34 Ibid.

35 The simulating system need not be an electronic computer. If the assumption of computational universality (see the next paragraph in the main text), on which this entire discussion is based, is correct, then the simulation could be performed using almost any objects, such as beer cans and string, or even something as simple as a classical three-body chaotic system, which is infinitely complex in its behaviour. Also, 'our' time and time in the simulating system need not be the same. The simulation could be much faster or much slower in its own time than our subjective experience of time within the simulation.

36 Barrow, 'Glitch'.

37 Paul Davies, 'A brief history of the multiverse', *New York Times*, 12 April 2003.

38 Martin Rees, 'In the matrix', *Edge* (www.edge.org), 15 September 2003.

9: Intelligent and Not So Intelligent Design

1 Augustine, *City of God*, XI, 4, 2, in *Basic Writings of St. Augustine*, edited by W. T. Oates (Random House, New York, 1948).

2 Aquinas is famous for his arguments for the existence of God, based on 'five ways' of reasoning. The five ways are contained in his *Summa Theologica*, edited by Timothy McDermott (Christian Classics Inc., Westminster, Md, 2000).

3 William Paley, *Natural Theology* (1802), in *Paley's Natural Theology with Illustrative Notes*, edited by H. Brougham and C. Bell (London, 1836), chapters 1 and 2.

4 Richard Dawkins, *The Blind Watchmaker* (Penguin, London, 1990).

5 Henry Drummond, *The Lowell Lectures on the Ascent of Man* (J. Pott & Co., New York, 1894), pp. 427–8.

6 The term seems to have been coined by C. A. Coulson in *Science and Christian Belief* (Fontana, London, 1958), although Drummond had already captured the basic idea in *The Ascent of Man*.

7 A useful video demonstrating the details, featuring Dan-Erik Nilsson, has been produced by WGBH Educational Foundation and Clear Blue Sky Productions, Inc., and can be found at http://www.pbs.org/wgbh/evolution/library/01/1/l_011_01.htm

8 Intelligent Design proponents are (for parochial US political reasons) frustratingly vague about the non-Darwinian mechanism whereby physical systems such as the bacterial flagellum acquire their design-like structure. It does not have to be an on-the-spot miracle, like a rabbit pulled out of a

hat, although that is apparently what their supporters prefer. There could be a design-like law of nature that operates over evolutionary timescales. To establish the meaningfulness of such a law, it is first necessary to provide a rigorous mathematical definition of 'design'. A heroic attempt at just that has been made by William Dembski; see his book *No Free Lunch* (Rowman & Littlefield, Lanham, Md, 2001).

9 A robust case for self-organization in biology is made by Stuart Kauffman in his book *At Home in the Universe* (Oxford University Press, Oxford, 1995).

10 Lee Smolin proposed a theory in which black holes create 'baby universes' which inherit laws from their 'parent universe', with some random variation. In this theory there is a sort of inheritance and variation, but no selection. Details can be found in his book *Life of the Cosmos* (Weidenfeld & Nicolson, London, 1997).

11 Christoph Schönborn, 'Finding design in nature', *New York Times*, 7 July 2005.

12 See, for example, Nelson Pike, *God and Timelessness* (Routledge & Kegan Paul, London, 1970).

13 E. W. Harrison, 'The natural selection of universes containing intelligent life', *Quarterly Journal of the Royal Astronomical Society*, vol. 36, no. 3 (1995), p. 193.

14 Remember, the landscape is not a physical place or region, but a space of possibilities – a parameter space. The super-being or super-civilization could create a universe physically close by, but a long way away in parameter space. If the universe containing this being or civilization were already optimal for life, we can imagine that it/they would choose to create baby universes at a similar location in the landscape, to make their product universes fit for life.

15 Olaf Stapledon, *The Star Maker* (Methuen, London, 1937).

16 Fred Hoyle, *The Intelligent Universe* (Michael Joseph, London, 1983), p. 249.

17 Andrei Linde, 'Stochastic approach to tunneling and baby universe formation', *Nuclear Physics*, vol. B372 (1992), p. 421.

18 Heinz Pagels, *The Dreams of Reason* (Bantam, New York, 1989), p. 156.

19 James Gardner, *Biocosm* (Inner Ocean Publishing, Maui, Hawaii, 2003), p. 178.

20 A clear discussion is given by Richard Swinburne, *The Coherence of Theism* (Clarendon Press, Oxford, 1977), part III.

21 That is, can a being that *exists* necessarily, is *good* necessarily, is *omnipotent* necessarily, and so on, also *not* create necessarily? Can a necessary being choose to not create?

22 Isaac Newton, who wrote more about theology than physics, used this argument. He reasoned that space and time at least are necessary because they emanate directly from God's necessary being. This may have been a factor in Newton's view that space and time are absolute, universal and unchangeable. Of course, we now know that this is wrong.

23 A good place to start is Keith Ward, *God: A Guide for the Perplexed* (Oneworld Publications, Oxford, 2005).

24 The unique, no-free-parameters theory is indifferent about whether there is only one representation of the universe or many. If there are many, they will be in identical quantum states – the postulated unique vacuum state of the theory. Because of the inherent uncertainty of quantum mechanics, this does not require the universes to be precise clones. So even the supposedly 'unique' universe theory is consistent with a limited form of multiverse.

25 The original idea for this analogy came from Carl Sagan, who described it in his novel *Contact* (Simon & Schuster, New York, 1985). It has been used in its present context by Rodney Holder in his book *God, The Multiverse and Everything* (Ashgate, Aldershot, 2004).

26 There is also a technical explanation, in terms of the foundations of mathematics and logic, of why a unique final theory is impossible. This has to do with what is known as Gödel's incompleteness theorem. For a recent discussion of this theorem, see, for example, Gregory Chaitin, *Meta Math! The Quest for Omega* (Pantheon Books, New York, 2005). It was partly in consideration of Gödel's theorem that Stephen Hawking, in a much publicized U-turn, recently repudiated the existence of a unique theory of everything.

27 Stephen Hawking, *A Brief History of Time* (Bantam, New York, 1988), p. 174.

28 Leibniz, who was a theist, considered this problem, and famously concluded that ours is the *best* of all possible worlds (for why would an all-good, perfect God create something less than best?). Leibniz's definition of 'best' refers not to maximum happiness for humans, but more abstractly to mathematical optimization: simplicity consistent with richness and diversity.

29 Anything that is logically self-consistent, I mean. A round square, for example, could not exist anywhere.

30 Tegmark was certainly not the first to suggest that all possible universes really exist. The idea was embraced, for example, by the Princeton philosopher David Lewis.

31 Max Tegmark, 'Parallel universes', *Scientific American* (May 2003), p. 31.

32 Ibid.

33 Benoît B. Mandelbrot, *The Fractal Geometry of Nature* (Freeman, New York, 1982).

34 Tegmark calls it the 'ultimate ensemble theory'.

35 See, for example, Chaitin, *Meta Math!*, p. 97.

36 The set of all sets that do not contain themselves is not in fact a set, according to the logical niceties of set theory.

37 Unless, that is, it can be demonstrated that there is a necessary being that is necessarily unique.

38 For example, one axiom states that any two points in space can be connected by a straight line.

39 This is perhaps a simplification. One may have evidential reasons for believing in a particular starting point. For example, support for a multiverse might come from evidence of variations of the 'constants' of nature. Support for God might come from religious experience or moral arguments.

40 Sometimes as 'the only one', but I have already pointed out the dubiousness of that claim.

41 Martin Gardner, *Are Universes Thicker Than Blackberries?* (Norton, New York, 2003), p. 3.

42 Richard Swinburne, *The Existence of God* (Oxford University Press, Oxford, 1979), chapter 5.

43 Richard Dawkins, 'The improbability of God', *Free Inquiry Magazine*, vol. 18, no. 3 (1998), p. 6.

44 Not the Tegmark multiverse: *that* is simple (well, maybe . . .).

10: How Come Existence?

1 Quoted by David Deutsch in *The Fabric of Reality* (Allen Lane, London, 1997), pp. 177–8.

2 Brandon Carter, 'Large number coincidences and the anthropic principle in cosmology', in *Confrontation of Cosmological Theories with Observational Data*, IAU Symposia No. 63, edited by M. S. Longair (Reidel, Dortrecht, 1974), p. 291; 'The anthropic principle and its implications for biological evolution', *Philosophical Transactions of the Royal Society of London A*, vol. 310 (1983), p. 347.

3 John Barrow and Frank Tipler, *The Anthropic Cosmological Principle* (Oxford University Press, Oxford, 1986).

4 Freeman Dyson, *Disturbing the Universe* (Harper & Row, New York, 1979), p. 250.

5 'Evolution's driving force', discussion between Robyn Williams and Simon

Conway Morris, ABC Radio National, 3 December 2005: http://www.abc.net.au/rn/science/ss/stories/s1517968.htm

6 Christian de Duve, *Vital Dust: Life as a Cosmic Imperative* (Basic Books, New York, 1995), p. 300.

7 Stuart Kauffman, *At Home in the Universe* (Oxford University Press, Oxford, 1995).

8 Deutsch, *The Fabric of Reality*, p. 181.

9 Ibid., p. 134. This statement is closely related to the Church–Turing thesis, the claim that defines the basis for the concept of a universal, or general-purpose, computer. Deutsch proposes elevating this thesis to the status of a fundamental principle of the universe.

10 Another example of an inconspicuous yet fundamental property of quantum systems is entanglement, where two or more particles remain subtly linked even though widely separated.

11 There is increasing evidence that some 'junk' DNA, although not part of the genetic coding system, may nevertheless play a role in the operation of the cell.

12 The crucial and basic distinction between the 'easy' and 'hard' problems of consciousness was first stressed by David Chalmers in a famous essay, 'Facing up to the problem of consciousness', *Journal of Consciousness Studies*, vol. 2 (1995), p. 200. Most, but not all, philosophers have since accepted this distinction as valid.

13 Daniel Dennett, *Consciousness Explained* (Little Brown, Boston, 1991).

14 See, for example, David Chalmers, *The Conscious Mind: the Search for a Fundamental Theory* (Oxford University Press, Oxford, 1997).

15 Just as Schrödinger's cat is seemingly in a state of 'suspended animation' in the absence of an observation, so the quantum universe as a whole remains suspended in a superposition of vastly many 'histories'. Readers who want to know more about the disappearance of time in quantum cosmology can find a detailed discussion in *The End of Time* by Julian Barbour (Weidenfeld & Nicolson, London, 2001).

16 Andrei Linde, 'Inflation, quantum cosmology and the anthropic principle', in *Science and Ultimate Reality*, edited by John Barrow, Paul Davies and Charles Harper (Cambridge University Press, Cambridge, 2004), p. 426.

17 Quoted by Tim Folger, 'Does the universe exist if we're not looking?' *Discover Magazine*, vol. 23, no. 6 (June 2002), p. 43.

18 Some suggestions for how this may be achieved have been made by Charles Lineweaver and myself: see P. C. W. Davies and Charles H. Lineweaver, 'Finding a second sample of life on Earth', *Astrobiology*, vol. 5, no. 2 (April 2005), p. 154.

19 Physicists often refer to this as a Boltzmann gas, after Ludwig Boltzmann, who studied how gases approach thermodynamic equilibrium.

20 I'm referring here to the macroscopic state, defined by averaging over many molecules, not to the micro-states in which the motions of individual molecules are specified.

21 This point is well recognized by scientists, and attempts have been made to provide a more precise definition of the elusive quality of 'organized complexity' that seems to characterize life. One promising definition, introduced by Charles Bennett of IBM, is in terms of the computational labour needed to describe the system. Bennett calls this the 'depth' of the system. A related but more physics-based definition of depth was proposed by Seth Lloyd and Heinz Pagels. A popular account of depth can be found in Murray Gell-Mann, *The Quark and the Jaguar* (Freeman, New York, 1994), pp. 100–105.

22 See, for example, Stuart Kauffman, *Investigations* (Oxford University Press, Oxford, 2002); or Eric Chaisson, *Epic of Evolution: Seven Ages of the Cosmos* (Columbia University Press, New York, 2005).

23 Even professional biologists are not immune from backsliding on this issue. In a recent article taking them to task, Charles Lineweaver highlights what he calls the 'planet of the apes fallacy'. See *Astrobiology*, vol. 5 no. 5 (2005), p. 658.

24 See, for example, Daniel Dennett, *Darwin's Dangerous Idea* (Simon & Schuster, New York, 1996).

25 An interesting case in point is the Gaia theory of life on Earth, according to which our planet's ecology, geology and climate form an interconnected dynamic feedback system in which Earth and its biosphere somehow cooperate to perpetuate life, for example by responding to external changes such as solar variability with compensating climatic changes. In this popular form, the Gaia theory looks decidedly teleological – Earth's biosphere responds to internal and external threats to secure its future – and it has been roundly criticized as such.

26 *Marx and Engels, Works,* vol. 40 (Moscow, 1929), p. 550.

27 Murray Gell-Mann, 'Nature conformable to herself', *Complexity*, vol. 1, no. 4 (1995), p. 1126.

28 Gell-Mann, *The Quark and the Jaguar*, p. 212. As an ironical aside, let me point out that if the extended version of the multiverse theory is considered (the one in which all possible laws are instantiated in a universe somewhere), then included within this multiverse there *must* be universes with teleological laws. One cannot banish teleological laws by fiat and at the same time argue that *all* possible laws are permitted in a universe somewhere. So if one embraces the extended multiverse theory, the question

then arises as to whether our universe is one of those that actually *has* teleological laws, or whether it hasn't, but is cunningly cooked up to mimic the genuine article. If universes with teleological laws exist, ours would be an excellent candidate. The universe certainly looks as if it possesses teleological features. Well, perhaps it *is* teleological!

29 Heinz Pagels, *Perfect Symmetry* (Michael Joseph, London, 1985), p. 347.

30 Časlav Bruckner and Anton Zeilinger, 'Information and fundamental elements of the structure of quantum theory', in *Time, Quantum and Information*, edited by Lutz Castell and Otfried Ischebeck (Springer-Verlag, Berlin, 2003), p. 323.

31 John Wheeler, 'On recognizing "law without law"', *American Journal of Physics*, vol. 51 (1983), p. 398.

32 This is not just the emergence of low-energy effective laws via symmetry-breaking, as discussed in Chapter 8. Wheeler proposes that *all* laws emerge from chaos after the origin of the universe.

33 John Wheeler, 'Information, physics, quantum: the search for links', in *Proceedings of the 3rd International Symposium on the Foundations of Quantum Mechanics*, Tokyo, 1989, p. 354.

34 John Wheeler, 'Frontiers of time', in *Problems in the Foundations of Physics*, edited by G. Toraldo di Francia (North-Holland, Amsterdam, 1979), p. 395.

35 This is a general statement, but in practice the bits are determined by quantum mechanics, in the form of discrete yes/no answers, such as whether an electron's spin is up or down.

36 John Wheeler, *At Home in the Universe* (AIP Press, New York, 1994), pp. 295–311. An attempt to build all of physics out of information has been made by B. Roy Frieden in *Physics From Fisher Information* (Cambridge University Press, Cambridge, 1998). For up-to-date comment on 'it from bit', see *Science and Ultimate Reality*, edited by John Barrow, Paul Davies and Charles Harper (Cambridge University Press, Cambridge, 2004), Part IV. See also Wheeler, 'Information, physics, quantum . . .'.

37 Two relevant papers by Rolf Landauer are 'Wanted: a physically possible theory of physics', *IEEE Spectrum*, vol. 4, no. 9 (1967), p. 105; and 'Computation and physics: Wheeler's meaning circuit?', *Foundations of Physics*, vol. 16, no. 6 (1986), p. 551.

38 Gregory Chaitin, *Meta Math! The Quest for Omega* (Pantheon Books, New York, 2005), p. 115.

39 Seth Lloyd's calculation is described in his paper 'Computational capacity of the universe', *Physical Review Letters*, vol. 88 (2002), p. 237901. See

also his book *The Computational Universe* (Random House, New York, 2006).

40 There may, however, be situations involving complex systems in which the limit of 10^{120} does matter. See P. C. W. Davies, 'Emergent biological principles and the computational resources of the universe', *Complexity*, vol. 10, no. 2 (2004), p. 1.

41 Paul Benioff, 'Towards a coherent theory of physics and mathematics', *Foundations of Physics*, vol. 32 (2002), p. 989.

42 Benioff's proposed consistency criterion is that the theory should maximally describe its own validity and sufficient strength.

43 Benioff, 'Towards a coherent theory . . .', p. 1005.

44 I have given a popular account in my book *About Time* (Viking, London, 1995).

45 F. Hoyle and J. V. Narlikar, *Direct Inter-particle Theories in Physics and Cosmology* (Freeman, San Francisco, 1974).

46 M. Gell-Mann and J. B. Hartle, 'Time symmetry and asymmetry in quantum mechanics and quantum cosmology', in *Physical Origins of Time Asymmetry*, edited by J. J. Halliwell, J. Pérez-Mercader and W. H. Zurek (Cambridge University Press, Cambridge, 1994), p. 311.

47 S.W. Hawking, 'The no boundary condition and the arrow of time', in *Physical Origins of Time Asymmetry*, edited by J. J. Halliwell, J. Pérez-Mercader and W. H. Zurek (Cambridge University Press, Cambridge, 1994), p. 346; Hawking subsequently retracted the idea.

48 Light rays can be bent by material obstacles, such as the edges of the slit. Thus photons do not always travel in precisely straight lines.

49 Quantum mechanics requires that all particles have a wave aspect. The two-slit experiment has, for example, been successfully carried out with electrons.

50 In a practical laboratory experiment the photon would take only nanoseconds to pass from the slits to the blind, and no human experimenter could make a decision so finely judged as to take place after the photon had traversed the slits but before it reached the blind. But this is a minor quibble. In principle one could make the distance to the image screen as long as one likes.

51 W. C. Wickes, C. O. Alley and O. Jakubowicz, 'A "delayed-choice" quantum mechanics experiment', in *Quantum Theory and Measurement*, edited by John A. Wheeler and Wojciech H. Zurek (Princeton University Press, Princeton, 1983), p. 457; T. Hellmuth, H. Walther, A. Zajonc and W. Schleich, 'Delayed-choice experiments in quantum interference', in *Physical Review A*, vol. 35 (1987), p. 2532.

52 There are lots of ingenious refinements to this scenario, and many actual experiments, including some in which the accomplice can make a record and then erase it. In all cases, no information can be sent back in time by this sort of arrangement.

53 A rather natural way of considering the delayed-choice experiment comes from the so-called transactional interpretation of quantum mechanics, due to John Cramer of the University of Washington (see *Reviews of Modern Physics*, vol. 58, (1986), p. 647). The essential idea is that a quantum event, such as the scattering of an electron or the decay of an atom, involves processes that go both forwards and backwards in time at the speed of light. If the transactional interpretation were applied to the universe as a whole, it might yield a self-consistent description. The challenge would then be to demonstrate that this description was unique.

54 Wheeler, 'Information, physics, quantum . . .', p. 354.

55 The concept of a self-explanatory loop is reflected in the ancient mystical symbol of the Ouroboros, represented as a snake eating its own tail.

56 Wheeler's more precise definition was 'a self-referential deductive axiomatic system' (see 'Information, physics quantum . . .', p. 357).

57 Barrow and Tipler, *The Anthropic Cosmological Principle*; Frank Tipler, *The Physics of Immortality* (Doubleday, New York, 1994).

58 'The anthropic universe', a documentary on the Australian Broadcasting Corporation's Radio National, *The Science Show*, 18 February 2006, produced by Martin Redfern and Pauline Newman. A transcript may be found at http://www.abc.net.au/rn/science/ss/stories/s1572643.htm

59 John Wheeler, 'World as a system self-synthesized by quantum networking', *IBM Journal of Research and Development*, vol. 32, no. 1 (1988), p. 4.

60 Quoted by Tim Folger, 'Does the universe exist if we're not looking?' *Discover*, vol. 23, no. 6 (2002), p. 48.

61 Another way to avoid paradoxes is to adopt the many-universes interpretation of quantum mechanics. See Deutsch, *The Fabric of Reality*, chapter 12.

62 If this skimpy discussion leaves the reader more confused than before, I can recommend my little book *How to Build a Time Machine* (Penguin, London, 2002) for more details.

63 Physicists will recognize this cumbersome description as what is technically termed a closed time-like world line.

64 J. R. Gott III and L.-X. Li, 'Can the universe create itself?' *Physical Review D*, vol. 58 (1998), p. 023501.

65 P. C. W. Davies, 'Closed time as an explanation of the black body background radiation', *Nature Physical Science*, vol. 240 (1972), p. 3.

66 Other scientists have had similar ideas. For example, Fred Hoyle concluded that, 'The Universe is seen as an inextricably linked loop . . . Everything exists at the courtesy of everything else.' *The Intelligent Universe*
(Michael Joseph, London, 1983), p. 248.

67 Wheeler arrived at a similar position. He insisted that the results of
quantum observations must *mean* something before the universe can be
said to be fully actualized. In this 'meaning circuit' (depicted in Figure 35,
p. 287), the physical world gives rise to 'observership' and 'meaning', while
observers and meaning loop back and give rise to the physical world. See,
for example, Wheeler, 'World as system . . .'.

68 Landauer, 'Computation and physics . . .'.

69 S. W. Hawking and T. Herzog, 'Populating the landscape: a top down
approach', hep-th/0602091. A popular account is Amanda Gefter, 'Mr.
Hawking's flexiverse', *New Scientist* (22 April 2006), p. 28.

70 Memes play the same role in human culture that genes play in genetics.
They may be, for example, habits, fashions or belief systems. Memes
replicate, spread within the community and compete.

Afterword: Ultimate Explanations

1 Broadcast in 1948 on the Third Programme of the BBC. Transcript
reprinted in Bertrand Russell, *Why I Am Not a Christian* (Allen & Unwin,
London, 1957), p. 155.

2 Neil A. Manson, 'Introduction', in *God and Design: the Teleological Argument and Modern Science*, edited by Neil Manson (Routledge, London,
2003), p. 18.

Bibliography

Some 'big picture' books about life, the universe and everything

Bill Bryson, *A Short History of Nearly Everything* (Broadway, New York, 2004)

Paul Davies, *The Mind of God* (Penguin, London, 1993)

David Deutsch, *The Fabric of Reality* (Allen Lane, London, 1997)

Timothy Ferris, *The Whole Shebang* (Simon & Schuster, New York, 1997)

James Gardner, *The Selfish Biocosm* (Inner Ocean Publishing, Maui, Hawaii, 2003)

Murray Gell-Mann, *The Quark and the Jaguar* (Freeman, New York, 1994)

Lawrence Krauss, *Atom: An Odyssey from the Big Bang to Life on Earth . . . and Beyond* (Little, Brown, 2001)

Seth Lloyd, *Programming the Universe: A Quantum Computer Scientist Takes on the Cosmos* (Random House, New York, 2005)

Roger Penrose, *The Road to Reality* (Jonathan Cape, London, 2004)

Lee Smolin, *The Life of the Cosmos* (Oxford University Press, Oxford, 1997)

Trinh Xuan Thuan, *The Secret Melody* (Templeton Foundation Press, London, 2005)

John Archibald Wheeler and Kenneth Ford, *Geons, Black Holes and Quantum Foam: A Life in Physics* (Norton, New York, 1998, revised edition 2000)

Stephen Wolfram, *A New Kind of Science* (Wolfram Media Inc., Champaign, Ill., 2002)

The unification of physics, subatomic particles, string/M theory, symmetry, etc.

John Barrow, *Theories of Everything* (Clarendon Press, Oxford, 1991)

Frank Close, *Lucifer's Legacy* (Oxford University Press, Oxford, 1999)

Frank Close, Michael Marten and Christine Sutton, *The Particle Explosion* (Oxford University Press, Oxford, 2002)

Brian Greene, *The Elegant Universe* (Vintage, New York, 2000)

Brian Greene, *The Fabric of the Cosmos* (Allen Lane, London, 2004)

Paul Halpern, *The Great Beyond: Higher Dimensions, Parallel Universes and the Extraordinary Search for a Theory of Everything* (Wiley, New York, 2004)

Michio Kaku, *Parallel Worlds* (Penguin, London, 2004)

Leon M. Lederman and Christopher T. Hill, *Symmetry and the Beautiful Universe* (Prometheus, New York, 2004)

Yuval Ne'eman and Yoram Kirsh, *The Particle Hunters* (Cambridge University Press, Cambridge, 1996)

Lisa Randall, *Warped Passages* (Allen Lane, London, 2005)

Michael Riordan, *The Hunting of the Quark* (Simon & Schuster, New York, 1987)

Steven Weinberg, *Dreams of a Final Theory* (Pantheon, New York, 1992)

Steven Weinberg, *The Discovery of Subatomic Particles* (Penguin, London, 1993)

Introductions to cosmology

Fred Adams and Greg Laughlin, *The Five Ages of the Universe* (Simon & Schuster, New York, 1999)

John Gribbin, *In Search of the Big Bang* (Penguin, London, 1999)

Alan Guth, *The Inflationary Universe: The Quest for a New Theory of Cosmic Origins* (Addison-Wesley, Reading, Mass., 1997)

Edward Harrison, *Cosmology: The Science of the Universe*, 2nd edition (Cambridge University Press, Cambridge, 2000)

Janna Levin, *How the Universe Got Its Spots* (Princeton University Press, Princeton, NJ, 2002)

Mario Livio, *The Accelerating Universe: Infinite Expansion, the Cosmological Constant, and the Beauty of the Cosmos* (Wiley, New York, 2000)

Dennis Overbye, *Lonely Hearts of the Cosmos* (HarperCollins, New York, 1991)

Simon Singh, *Big Bang: The Origin of the Universe* (Fourth Estate, London, 2004)

Books including discussions of multiverse ideas and 'anthropic' selection

Fred Adams, *Our Living Multiverse* (Pi Press, New York, 2004)

John Barrow and Frank Tipler, *The Anthropic Cosmological Principle* (Oxford University Press, Oxford, 1986)

Paul Davies, *The Accidental Universe* (Cambridge University Press, Cambridge, 1982)

John Gribbin and Martin Rees, *Cosmic Coincidences: Dark Matter, Mankind, and Anthropic Cosmology* (Bantam, New York, 1989)

John Leslie, *Universes* (Routledge, London, 1989)

Martin Rees, *Before the Beginning: Our Universe and Others* (Simon & Schuster, London, 1997)

Martin Rees, *Just Six Numbers: The Deep Forces that Shape the Universe* (Basic Books, New York, 1999)

Martin Rees, *Our Cosmic Habitat* (Princeton University Press, Princeton, NJ, 2001)

Leonard Susskind, *The Cosmic Landscape: String Theory and the Illusion of Intelligent Design* (Little Brown, New York, 2005)

Alex Vilenkin, *Many Worlds in One: The Search for Other Universes* (Hill and Wang, New York, 2006).

Crossover into religion and theology

Wim Drees, *Beyond the Big Bang: Quantum Cosmology and God* (Open Court, La Salle, Ill., 1990)

Rodney Holder, *God, the Multiverse and Everything* (Ashgate, Aldershot, 2004)

Fred Hoyle, *The Intelligent Universe* (Michael Joseph, London, 1983)

Neil Manson (editor), *God and Design* (Routledge, London, 2003)

John Polkinghorne, *Science and Providence* (Templeton Foundation Press, London, 2005)

Joel Primack and Nancy Abrams, *The View from the Centre of the Universe* (Fourth Estate, London, 2006)

Russell Stannard, *Science and the Renewal of Belief* (Templeton Foundation Press, London, 2005)

Victor J. Stenger, *Has Science Found God? The Latest Results in the Search for Purpose in the Universe* (Prometheus Books, New York, 2003)

Frank Tipler, *The Physics of Immortality* (Doubleday, New York, 1994)

Index

References to diagrams are given in *italics*.

He just wanted a decent book to read ...

Not too much to ask, is it? It was in 1935 when Allen Lane, Managing Director of Bodley Head Publishers, stood on a platform at Exeter railway station looking for something good to read on his journey back to London. His choice was limited to popular magazines and poor-quality paperbacks – the same choice faced every day by the vast majority of readers, few of whom could afford hardbacks. Lane's disappointment and subsequent anger at the range of books generally available led him to found a company – and change the world.

'We believed in the existence in this country of a vast reading public for intelligent books at a low price, and staked everything on it'
Sir Allen Lane, 1902–1970, founder of Penguin Books

The quality paperback had arrived – and not just in bookshops. Lane was adamant that his Penguins should appear in chain stores and tobacconists, and should cost no more than a packet of cigarettes.

Reading habits (and cigarette prices) have changed since 1935, but Penguin still believes in publishing the best books for everybody to enjoy. We still believe that good design costs no more than bad design, and we still believe that quality books published passionately and responsibly make the world a better place.

So wherever you see the little bird – whether it's on a piece of prize-winning literary fiction or a celebrity autobiography, political tour de force or historical masterpiece, a serial-killer thriller, reference book, world classic or a piece of pure escapism – you can bet that it represents the very best that the genre has to offer.

Whatever you like to read – trust Penguin.